SECOND EDITION

RADIATION
AND
RADIOACTIVITY
ON
EARTH AND BEYOND

Ivan G. Draganić, D. Sc.
Senior Scientist
The Boris Kidrič Institute
Vinča, Yugoslavia

Zorica D. Draganić, D. Sc.
Senior Scientist
The Boris Kidrič Institute
Vinča, Yugoslavia

Jean-Pierre Adloff, D. Sc.
Titular Professor
Université Louis Pasteur
Strasbourg, France

CRC Press
Boca Raton   Ann Arbor   London   Tokyo

Library of Congress Cataloging-in-Publication Data

Draganić, Ivan G.
    Radiation and radioactivity on earth and beyond / authors, Ivan G.
Draganić, Zorica D. Draganić, Jean-Pierre Adloff. -- 2nd ed.
        p.   cm.
    Includes bibliographical references and index.
    ISBN 0-8493-8675-6
    1. Ionizing radiation.   2. Radioactivity.   3. Earth sciences.
4. Astrophysics.   I. Draganić, Zorica D.   II. Adloff, J. P. (Jean
Pierre)   III. Title.
    QC795.D73   1993
    539-7--dc20   92-46189
                                                                    CIP

    Direct all inquiries to CRC Press, Inc., 2000 Corporate Blvd., N.W., Boca
Raton, Florida 33431.

International Standard Book Number 0-8493-8675-6

Library of Congress Card Number 92-46189

Printed in the United States of America   2 3 4 5 6 7 8 9 0

Printed on acid-free paper

Preface to the Second Edition

Since the first edition appeared there have been no scientific and technological events in the fields of radiation and radioactivity which would obligate the authors to make essential changes. However, there was a major geopolitical event. The nuclear arms race between the two nuclear superpowers has subsided and even the dismantling of nuclear weapons has been undertaken. But numerous concerns still remain due to the political and social instability in the world. The scope of some of the revisions took into account this and other facts in the hope of making the content of the book current and relevant.

Active participation in the preparation of this edition was interrupted by the death of one of the authors. Dr. Zorica D. Draganić died in October 1991. We greatly miss her gentle, knowledgeable, and stimulating participation. We have tried to complete the task in the spirit in which she would approach it.

Ivan G. Draganić

Jean-Pierre Adloff

Preface

Ionizing radiation and radioactivity were discovered at the end of the last century. Since then, they have continued to make an impact on various fields of research, including not only fundamental studies ranging from the structure of matter at the subatomic level up to cosmic space and the evolution of the universe, but also numerous applications in everyday life, industry, medicine, and energy production. The achievements made have profoundly marked our civilization.

However, radiation and radioactivity are also responsible for many apprehensions that have dominated public opinion during the latter half of this century. The layman has an instinctive fear of radiation that is undetectable by his senses, and which he knows to be harmful. He is also aware of the permanent risk of accidents such as that of Chernobyl, with the devastating long-term consequences which affect the population, the environment and national economy. The horrendous threat of nuclear war is still a menace to the very existence of our civilization, despite recent and encouraging steps toward armament limitations.

Clumsy official statements relating to the safety of nuclear installations or reports of difficulties with radioactive waste disposal in which harmful consequences are minimized often stimulate heated debate. Public opinion is strongly polarized. The opponents of nuclear power programs are becoming increasingly sensitive to all applications of radioactivity and radiation and frequently contend that accounts of accidents in the press merely reveal the tip of an iceberg which conceals its reality behind commercial or military secrets. According to this opinion, the harmful effects are overwhelmingly greater than the benefits, and the global risks so large that not only the production of nuclear energy but also all applications of radiation and radioactivity should be abandoned. In contrast, supporters insist that the mass media often exaggerate the gravity of accidents, thereby distorting the public's viewpoint on nuclear energy. It is argued that in the near future, until alternative energy sources are developed, the world needs nuclear power for meeting its energy requirements; it is further maintained that many applications of radioactivity and radiation, for example in medical diagnosis and therapy, clearly present more benefits than risks and should therefore be used until corresponding nonradioactive techniques are accessible.

In presenting the facts, we have tried neither to minimize nor exaggerate the risks and benefits; at the same time, we have not avoided commenting on important uncertainties.

This book is written as an introductory text for those who have a basic knowledge in the natural sciences and a general interest in radiation, radioactivity, and nuclear energy on the Earth and in cosmic space.

In this hopefully objective approach we have also taken into account that public opinion largely ignores the fact that radiation, radioactivity, and nuclear energy are primarily natural phenomena which play an important role in the life of the cosmos, the solar system and our planet. It is not generally known that nuclear reactors were already operating on the Precambrian Earth at a time when primitive forms of life were emerging and the evolutionary process began. Natural sources of ionizing radiation, both of terrestrial and extraterrestrial origin, have been a part of our environment from the very beginning. Radioactive nuclides are incorporated in the human body. Some of them came into being already thousands of millions of years ago in the course of violent nuclear events in stars and interstellar space. After aeons of recycling in cosmic space and on Earth, they became a natural constituent of our body, thus confirming the poetical saying that we are "composed of the dust of the stars"!

In choosing and arranging the subject matter, we have necessarily dealt with various fields such as nuclear physics and chemistry, nuclear technology, radiochemistry, hot atom chemistry, radiation physics, and radiation chemistry. The main intention was to present a concise and general picture, avoiding as much as possible a textbook presentation of achievements. The detailed theoretical or technical treatment of the subject has been reduced to a minimum. For the interested reader seeking further information, some literature references are given at the end of each chapter. It is hoped that the glossary of scientific and technical terms at the end of the book will be helpful to readers who may be unfamiliar with various topics in this interdisciplinary approach to radiation and radioactivity.

We are grateful for helpful advice from colleagues who were kind enough to read the manuscript: Z. Dizdar and V. Ajdačić (Nuclear Sciences Institute, Vinča, Yugoslavia), S. Ribnikar and Z. Marić (University of Beograd, Yugoslavia), and Hiroshi Mizutani (Mitsubishi-Kasei Institute of Life Sciences, Tokyo, Japan). We are indebted to John MacCordick (Centre de Recherches Nucléaires, Strasbourg, France) for linguistic appreciation and critical reading of the manuscript. Whatever errors remain are, of course, the responsibility of the authors.

We wish to thank T. Ružičić (Beograd, Yugoslavia) for his assiduity in preparing the illustrations, G. Jaboulay (Strasbourg, France) for typing many parts of the text, and M. Adloff (Strasbourg, France) for the competence and meticulous care with which she helped to prepare the final version of the manuscript for the publisher.

During the major part of this work, two of the authors (I.G.D. and Z.D.D.) were on leave from their home institute (Vinča, Yugoslavia), as visiting scientists at the Universidad Nacional Autonoma de Mexico (CEN-UNAM, Ciudad Mexico, Mexico) and at RISÖ National Laboratory, Accelerator Department (Roskilde, Denmark). They are grateful both to the host institutions and to their directors, Marcos Rosenbaum (UNAM) and K. Sehested (RISÖ), for providing excellent library facilities and a stimulating atmosphere for work on the manuscript.

Ivan and Zorica Draganić, Vinča-Yugoslavia
Jean-Pierre Adloff, Strasbourg-France
September 1989

Authors

Ivan G. Draganić is a senior scientist at the Nuclear Sciences Institute (Vinča, Yugoslavia), where he has worked with ionizing radiation since 1950. Dr. Draganić received a D.Sc. from the Sorbonne, Paris and has been a visiting scientist at the Centre d'Études Nucléaires de Saclay (France), the Risø National Laboratory (Denmark), the Brookhaven National Laboratory (U.S.), and the Universidad Nacional Autonoma de Mexico (Mexico). He is a member of working groups and field-expert of the International Atomic Energy Agency as well as author and co-author of about 150 technical and scientific articles on the dosimetry of ionizing radiation (nuclear reactors), radiation chemistry (aqueous systems), and evolutionary processes (early Earth, cometary nuclei). In addition, Dr. Draganić is the author, co-author, and editor of more than a dozen books including *Introduction to Radiochemistry, Radiochemical Procedures, Radioactive Isotopes and Radiation,* and *The Radiation Chemistry of Water.* In 1973 he shared the National Award for research with his wife, Dr. Zorica D. Draganić.

Zorica D. Draganić was a senior scientist at the Nuclear Sciences Institute (Vinča, Yugoslavia), where she had worked with ionizing radiation since 1950. Dr. Draganić received a D.Sc. from the University of Belgrade, Yugoslavia and has been a visiting scientist at the Centre d'Études Nucléaires de Saclay (France), the Brookhaven National Laboratory (U.S.), the Universidad Nacional Autonoma de Mexico (Mexico), and the Risø National Laboratory (Denmark). She was author and co-author of about 90 scientific articles on radiation chemistry (water, aqueous solutions) and radiation chemical aspects of chemical evolution processes (Precambrian Earth, comet nuclei). In addition, Dr. Draganić was co-author of several books including *The Radiation Chemistry of Water*. She shared with her husband, Dr. I. G. Draganić, the National Award for Research in 1973.

Jean-Pierre Adloff is Titular Professor of Nuclear Chemistry at the University of Strasbourg (France) and Director of the Laboratory of Nuclear Chemistry, Nuclear Research Center, Strasbourg. He received a D.Sc. from the University of Strasbourg where he had prepared his doctorate with the late Professor Marguerite Perey, discoverer of the element francium and former assistant of Marie Curie. Dr. Adloff is the author and co-author of about 100 articles on radiochemistry, radioanalytical chemistry, Mössbauer spectroscopy, and hot atom chemistry and is a committee member of the International Symposia on Hot Atom Chemistry and the International Conference in Nuclear and Radiochemistry. He is a participant in French scientific cooperation programs in Brazil, Mexico, and Paraguay and is a UNESCO expert in Brazil and an IAEA expert in Thailand, Ecuador, and Kenya. Professor Adloff is the Co-editor of *Radiochimica Acta*.

Contents

Chapter 10
Perspectives

This chapter concerns people and events in a story of radiation, radioactivity, and nuclear energy that covers almost a century.

The term radiation is generally used to denote the emission of any type of ray or particle from a source. If the radiation has sufficient energy to produce ions by either direct or indirect mean, it is called ionizing radiation. This book deals almost entirely with radiation of this kind. It originates in nuclear reactions and processes, appears in the form of cosmic rays, and is produced in laboratories by special devices.

Since the discovery of X-rays by Roentgen in 1895 and the identification of radiation from radioactive substances (alpha [α]- and beta [β]-particles and gamma [γ]-rays), the inventory of radiation is gradually being extended. Dozens of types of nuclear and subnuclear particles are presently being produced artificially in laboratories at energies up to 10^9 electronvolts or more, while cosmic radiation provides a variety of entities at energies up to 10^{20} electronvolts.

Radioactivity is a property of nuclei to undergo a spontaneous disintegration accompanied by emission of radiations. The first radioelements were discovered by Pierre and Marie Curie in 1898. At present, about 80 various radionuclides are known to occur in nature. Since the discovery of artificial radioactivity by Irène Curie and Frédéric Joliot in 1934, more than 2500 radioactive nuclides have been produced by man.

Energy released in a nuclear reaction involving the conversion of mass into energy is called nuclear energy. The fission of the uranium nucleus was discovered in 1939 and the first man-made nuclear reactor was put into operation in the United States in 1942, in Chicago. At the present time, there are more than 300 experimental nuclear reactors operating across the world and about 420 nuclear power plants furnish nearly 12 percent of the global output of electricity. Our era is also characterized by a prolific stockpile of nuclear weapons. Within only 4 decades following the explosion of the first atomic bomb in the desert of New Mexico (U.S.) in 1945, the world's nuclear armament has been built up to a point sufficient to destroy not only our civilization, but also much of the Earth's capacity to support life.

A Story That Covers Almost A Century

THE PENETRATING INVISIBLE RADIATION

The nuclear age began in 1896 with the discovery of radioactivity by Antoine-Henry Becquerel (1852—1908) (Figure 1). It followed, by 3 months, the discovery of X-rays by Konrad Wilhelm Roentgen (1845—1923). These invisible rays traveled over a distance of several meters, and were emitted when a gas at low pressure was excited by an electric discharge. Simultaneously, a green luminescence was observed on the walls of the glass tube containing the gas.

Becquerel thought that there should be some kind of relationship between the luminescence and the emission of penetrating radiation, and that a material which becomes luminescent after being exposed to sunlight might also emit X-rays. He knew that some uranium salts show strong fluorescence when exposed to sunlight and he decided to verify whether they also emit the invisible penetrating radiation. The radiation was detected by the blackening of a photographic plate in a very simple experiment: the sensitive plate was wrapped in black paper, a crystal of the uranium salt placed on it, and the whole exposed to sunlight. In a fortuitous but crucial observation, Becquerel noticed that the

FIGURE 1. Henri Becquerel (1852 — 1908).

sunlight and the luminescence were really not necessary for blackening of the photographic plate. He concluded that the blackening occurs from an intrinsic property of the uranium salt and he prepared a report on "the invisible radiation emitted by phosphorescent substances" for the weekly session of the Academy of Sciences in Paris.

Becquerel presented the report on March 2 (1896), a date which is usually considered as the birthday of radioactivity and nuclear science. In further experiments Becquerel found that the radiations were emitted by the chemical element uranium and that they shared several properties of the X-rays, such as the ability to discharge an electroscope.

RADIOACTIVITY AND THE CURIES

Two years after Becquerel's discovery of radiation emitted by uranium, Gerhart Carl Schmidt (1865—1949) reported that thorium, another heavy chemical element, could also discharge an electroscope and blacken a photographic plate. The same result was obtained independently by a young Polish woman, Marya Sklodowska (1867—1934), whose civil status changed in 1895 to Marie Curie. She became the most prominent figure in the history of "radioactivity", the name she gave to the new phenomenon of spontaneous emission of invisible radiation.

In 1882, her husband Pierre Curie (1859—1906) had discovered piezoelectricity, a phenomenon related to the appearance of electric charges when a quartz foil is subjected to pressure. He conceived a simple but very effective device to measure the electric charge. Even small charges, such as those resulting from radiation emitted by uranium, could be measured accurately. The device known as the piezoelectric quartz electrometer was used by generations of radiochemists, up to the advent of the Geiger-Müller counter and simple electronic tubes (Figure 2).

Marie Curie chose the study of rays emitted by uranium and thorium as a subject for her doctoral thesis. With the aid of the quartz electrometer, she also undertook precise measurements of every mineral and chemical compound she could find. In her first publication (April 1898), she noted that potassium salts are weakly radioactive, and further reported that two uranium minerals are more radioactive than uranium itself. The latter finding implied that the minerals pitchblende and chalcolite must contain some unknown constituent which was more "active" than uranium.

Together with Gustave Bémont (1867—1932), the Curies undertook the chemical processing of pitchblende. The procedure was in no way innovative, but the quartz electrometer enabled a simple and reliable control of the radioactivity after each separation step. In July 1898, they announced the discovery of a new element whose properties were similar to those of bismuth. It was named polonium in honor of Marie's native country.

The Curies soon discovered a second radioactive element, which resembled barium and for which they proposed the name radium. The optical spectrum of the new element showed characteristic lines, but final proof for its existence required an accurate determination of atomic weight.

Too many "discoveries" of new chemical elements had already been reported toward the end of the 19th century and the scientific

FIGURE 2. Pierre Curie (1859 — 1906) and Marie Curie (1867 — 1934).

community had become skeptical and cautious. Any claim had to be supported by "visible proofs". In addition to the establishment of characteristic spectral lines, such proofs comprised exhibition in a test tube and an unequivocal determination of the atomic weight.

The latter required a weighable quantity of radium, which the Curies extracted tediously from several tons of pitchblende residues remaining from the separation of uranium used for the manufacture of colored ceramic and glass. Finally, 100 milligrams of radium chloride was recovered after 4 years of exhausting physical labor in a miserable and unheated open shed. But this tiny amount was sufficient to provide visible evidence for the unquestionable existence of a new radioactive element. In 1903, the Curies shared with Becquerel the Nobel Prize in physics. Elemental radium in the metallic state was obtained by Marie Curie in 1910 and the following year this achievement brought her a second Nobel Prize, this time in chemistry. Marie Curie was the first person to obtain this prestigious award twice.

THE NATURE OF RADIOACTIVITY, RUTHERFORD AND SODDY

A common property of X-rays and uranium rays is their ability to ionize a gas. In 1899, Ernest Rutherford (1871—1937) began a systematic study of radiation emitted by uranium and examined the ionization under various conditions. He soon established that this radiation consists of two different types of particles. One of these, which he named α-particles (α-rays), produced a large number of ions, but was completely stopped by a thin sheet of glass, aluminum, or paper; these particles were identified as positive ions of the element helium. The other type of radiation was more penetrating but produced fewer ions. Rutherford called this form β-radiation (β-rays), which was found to consist of electrons, i.e., negatively charged particles. Since α- and β-rays bear electric charges, they are affected by electric and magnetic fields. This is not the case, however, for the third type of radiation, the γ-rays, which are electromagnetic waves like X-rays. γ-Radiation produces little ionization and travels over greater distances (Figure 3).

The years following the discovery of radium brought many, frequently confusing, observations. Several new "activities" were described but could not be ascribed to a specific element. One of the most important observations was that the radioactivity of a substance does not necessarily continue indefinitely; for a given substance, the activity decreases within a specific time scale.

Rutherford discovered that thorium releases a radioactive gas, which he called "emanation". This finding was useful in explaining the instability of radioactive species. Together with Frederick Soddy (1877—1956), Rutherford formulated a revolutionary idea on the nature of radioactivity at the beginning of 1903. It was recognized that radioactive decay is a process within the atom and involves its spontaneous transformation, i.e., the atom changes its chemical identity, and is "transmuted" into the atom of another element.

FITTING RADIOELEMENTS AND ISOTOPES INTO MENDELEEV'S PERIODIC TABLE

The flourishing field of radioactivity left many scientists rather indifferent, possibly because of the complicated or confusing cookbook type of procedures described by many chemists in their publications.

However, the chemists had their own troubles. The problem was how to fit the multitude of new radioactive species into the

FIGURE 3. Ernest Rutherford (1871 — 1937).

periodic table of chemical elements, as set up in 1869 by Dmitri Ivanovich Mendeleev (1834—1907) (Figure 4). Mendeleev's idea was that the properties of the elements depend in a periodic manner on their atomic weight. In such an arrangement, elements which have similar properties recur at regular intervals and fall into related groups. Mendeleev was able to predict the existence and properties of hitherto undiscovered elements by means of his original table, and subsequently the periodic law was extremely useful in explaining and correlating the properties of elements. In 1895, the table listed 66 elements, together with at least 10 others whose position was still uncertain. A disturbing fact was the incompatibility of several known atomic weights with the attractive and simple hypothesis of William Prout (1785—1850) that the atoms of chemical elements are composed of hydrogen atoms. The advance in experimental procedures used in the determination of atomic weights has shown that more and more data differ substantially from the values expected for a multiple of the atomic weight of hydrogen.

FIGURE 4. Dmitri Ivanovich Mendeleev (1834 — 1907).

By the end of 1911, three dozen new radioactive species had been definitely identified and characterized by their radiation and the kinetics of their decay, although room remained for only 14 further elements in the periodic table. The puzzle was progressively solved when it was recognized that one radioactive entity could be transformed into another and that a certain genetic relationship existed within a radioactive family. After a great deal of confusion and controversy it was found that all known radioactive elements could be assigned to one of three radioactive families. Two of these had uranium as the parent element and one was headed by thorium; the final product in all cases was lead.

The next task was to force the members of radioactive families into the vacancies of the periodic chart. Radium was easy to accommodate on the basis of its chemical analogy with barium and its known atomic weight. The three forms of emanation, which behaved like the inert gas argon, also had definitive positions.

For the five remaining spaces between bismuth and uranium there were still too many candidates. The idea emerged that some of the radioactive substances might represent varieties of the same element. The way to this hypothesis was opened in 1913 when Frederick Soddy and Kasimir Fajans (1887–1975) suggested that the lead which appears as a stable end product of the three radioactive families, and the practically inseparable ordinary lead of different atomic weight, occupy the same place in the periodic table. Some of the radioactive species also had similar, if not identical, chemical properties and thus belonged together. Their characteristic radioactive behavior indicated, however, that they are distinguishable as different substances.

A great step forward was made when Fajans announced the "displacement laws" in 1912. According to this scheme, emission of an α-particle is accompanied by a shift of the radioactive element in the periodic table by two places from right to left in the horizontal row, whereas β-emission results in a shift of one position from left to right in the horizontal row. Fajans was able to fit all known radioactive elements into the periodic table and assign their atomic weights. The valid prediction of chemical properties was maintained.

It was now firmly established that several radioactive elements of different origin, with different radioactive properties and atomic weights, could share the same place in the periodic table. Soddy coined the word "isotope", from Greek "in the same place". This concept resurrected Prout's image of atoms as multiaggregates of hydrogen atoms, but in a different way: the atomic weights of isotopes (not of atoms) should be the whole number multiple of the atomic weight of hydrogen. Thus, there is no need for separate positions in the periodic table for new radioactive isotopes, and the periodicity of the table is preserved. Soddy was awarded the Nobel Prize in chemistry in 1921.

RADIOACTIVITY AND PHYSICS

The new field of radioactivity was considered by physicists as a spectacular sorting out of new radioactive elements by methods which reposed primarily on the chemist's skill. Apart from this, however, they were not particularly impressed.

In fact, they were facing a major revolution of their own. In 1905, Albert Einstein (1879—1955) had published the theory of relativity, in which the equivalence of mass and energy was stated in what is now one of the most famous equations: $E = mc^2$, where E represents energy, m is the mass, and c is the speed of light. A few years before (1900), Max Planck (1858—1947) had introduced the concept that on the atomic scale energy can be absorbed or emitted only stepwise as multiples of a "quantum". This notion of the discontinuity of energy enabled Niels Bohr (1885—1962) to establish in 1913 a model of the hydrogen atom consisting of an electron revolving in one of several possible orbits around a central nucleus. The orbits are discrete and the electron can only "jump" between them after absorbing or releasing a quantum of energy.

The discovery of the photoelectric effect by Philipp Lenard (1862–1947), who received the Nobel Prize for physics in 1905, and the explanation of the phenomenon by Albert Einstein (for which he received the 1921 Nobel Prize in physics) have shown that light may act not only as a wave, but also as corpuscle. The wave-corpuscle dualism received further support in 1923 from Louis de Broglie (1892—1987), who assigned to each corpuscle a characteristic wavelength depending on its mass and velocity.

These findings had no direct impact on the progress of research in radioactivity in the 1920's, but played an important role in explaining the structure of the atom. Rutherford had experimentally shown in 1911 that the positive charge in the atom and most of its mass must be concentrated in a very small region, recognized as the nucleus. In the course of experiments in which he bombarded nitrogen with α-particles, Rutherford noticed the appearance of hydrogen atoms. He concluded that the hydrogen atom released by the target nucleus of nitrogen was a positive ion and he called it a *proton* (from Greek "the first"). The atom was regarded as an assembly consisting of a nucleus made up of protons and surrounded by electrons. This concept remained until the discovery of the neutron in 1932 by James Chadwick (1891—1974). From then on, the basic structure of the atom was established: a nucleus composed of protons and neutrons, surrounded by electrons.

ARTIFICIAL RADIOACTIVITY AND THE JOLIOT-CURIE TEAM

The first man-made "nuclear reaction" was realized by Rutherford in 1919. He bombarded nitrogen atoms with α-particles and the nitrogen atoms were transmuted into oxygen. For the first time, the alchemist's dream became — in principle — a reality: transmutation had occurred by man's intervention. The reaction yield was very low and it was obvious that for further study much more intense sources of α-particles were required. Many α-emitters among natural radioelements were available, but the only one that was feasible for the preparation of intense sources was polonium. Obviously, the best place to prepare large amounts of this element was the "Institut du Radium" in Paris, founded by the discoverer of the element. In 1925, Marie Curie assigned this task to her new assistant Frédéric Joliot (1900—1958). The following year, Joliot married Irène Curie (1897—1956), Marie's daughter and a close collaborator. The Joliot-Curie couple was soon to reach fame equal to that of Marie and Pierre Curie.

After Joliot had submitted his doctoral thesis on the electro-chemistry of polonium, he continued the work with his wife and together they made the strongest polonium source available at that time. They exposed certain light elements to the intense beam of α-particles and in January 1934 reported that some of them, like aluminum and boron, could be made artificially radioactive. It was also reported that the respectively formed isotopes, radio-phosphorus and radionitrogen, obey a law of exponential decay in the same way as do naturally occurring radioactive elements. The first man-made radioelements emitted a new type of radiation consisting of a positively charged electron, the positron. This particle had been predicted by Paul Dirac and until then its existence was confirmed only in cosmic rays. For the discovery of artificial radioactivity Joliot-Curie received the 1935 Nobel Prize for chemistry (Figure 5).

The discovery made an enormous impact on nuclear chemists and the number of synthesized radionuclides very rapidly expanded, soon reaching several hundred. At the present time, over 2500 synthetic radionuclides are known.

Two other events also contributed to the proliferation of artificial radioactive isotopes. One was the discovery of the neutron as an efficient projectile for bombardment and the availability of strong neutron sources. The second event was the development of accelerators producing ions with energies sufficient to enter into nuclei and induce nuclear transformations.

Devices for artificial acceleration of charged particles were first developed in the 1930s. Van de Graaff conceived in 1931 an electrostatic high-voltage generator. The following year, J.D. Cockroft (1897—1967) and E.T.S. Walton developed a high-voltage multiplier and used it for the first nuclear transformation by means of accelerated particles in the laboratory. In 1931, E.O. Lawrence (1901—1958) constructed the first cyclotron.

URANIUM FISSION

Shortly after Joliot-Curie's discovery of artificial radioactivity induced by α-particles, Enrico Fermi (1901—1954) decided to study transmutations by using the newly discovered neutral particles as projectiles. He suspected that neutrons could penetrate more easily into the nucleus than the positively charged α-

FIGURE 5. Iréne Curie (1897— 1956) and Frédéric Joliot (1900 — 1958).

rays. Fermi bombarded systematically all the known chemical elements in increasing order of atomic weight. In the spring of 1934, after more than 2 years of work involving the discovery of about 40 new radioactive nuclides, the moment came for uranium to be used as a target nucleus. This was the heaviest element known at that time and Fermi expected that the reaction products in this case might include a transuranium element. Indeed, several radioactive species were found in the irradiated samples and they were attributed to the first synthetic transuranium elements which so far were unknown on Earth.

These assignments were made because the chemical behavior of the induced radioactivities was found to be different from that of all known elements close to uranium in the periodic system of the elements. The hypothesis was also justified by the fact that the hitherto observed radioactive decays involving emission of β-particles were known to lead to a chemical element situated on the right side of the target element in the periodic table, i.e., toward the hypothetical transuranium species in the case of uranium.

Today we know that most of the radioelements formed in neutron-irradiated uranium have chemical properties in common with those of elements in the middle part of the periodic table, but this was not evident until further thrilling work had been accomplished. Even the scientists of the old guard of radiochemists like Otto Hahn (1879—1968), Liese Meitner (1878—1968), and their collaborator Fritz Strassman (1902—1980) reported in 1937 that the exposure of uranium to neutrons leads to radioactive species beyond uranium. They were forced to revise their opinion after Irène Joliot-Curie and her collaborator Pavle Savić (born 1909) published their results in 1938. These authors discovered a radioactive species with a half-life of 3.5 hours, which, after careful examination, was found to behave like the element lanthanum. This was a very unexpected result, since the atom of lanthanum is only half as large as that of uranium and is far removed from the latter in the periodic table. In fact, it really was lanthanum and Joliot-Curie and Savić were within a hair's breadth of discovering nuclear fission. However, owing to difficulty in the interpretation of residual radioactivity in the irradiated sample (activity they considered as an indication of some subtle difference between the chemical behavior of the 3.5 hour substance and that of lanthanum), they did not rule out the possibility that some peculiar transuranium species had been formed.

The work of Joliot-Curie and Savić had a profound effect on Otto Hahn. Initially, the latter was skeptical and, as he said to Frédéric Joliot, he "decided to repeat the experiments despite his great respect for Irène". Eventually, Hahn and Strassman confirmed the presence of radioactive species which behaved as typical chemical elements in the middle of the periodic table, namely, barium and lanthanum. When they presented their results in January 1939 in *Naturwissenschaften*, a German scientific journal, Hahn and Strassman pointed out that they were quite aware that their finding was in contradiction with all previous experience in nuclear physics.

The explanation of this strange behavior of neutron-irradiated uranium came only a few days after the article of Hahn and Strassman appeared. It was published in the English journal *Nature*, in a paper that was drafted over a long-distance telephone call by Liese Meitner and her nephew Otto Robert Frisch (1904–1975). They considered this new type of nuclear reaction as a splitting of the uranium nucleus into two smaller fragments, an event which they named "fission" by analogy with cell division in biology. They pointed out that the energy liberated in this process (200 million electronvolts) was much higher than that encountered in any previously known reaction, and was due to the conversion of mass into energy according to Einstein's relation. Accompanying their paper was a second note containing Frisch's experimental results that substantiated the very high energy of fission fragments.

Independently of Meitner and Frisch, and at the same time he received the issue of *Naturwissenschaften* containing Hahn and Strassmann's paper, Frédéric Joliot performed an experiment which showed that fission was responsible for the peculiar results with uranium and that the fission fragments had high energy. But it was not only the liberation of large amounts of energy that made fission so important. The fact that two or three neutrons were released simultaneously in the process was even more significant, and this was pointed out soon afterwards by Joliot and his collaborators as well as by several other nuclear physicists. It was also found that only the rare isotope ^{235}U underwent fission.

For his work with neutrons and the production of new radioactive isotopes, Enrico Fermi was awarded the Nobel Prize for physics in 1938. He also made several other discoveries which were important for nuclear energy and he occupied a key position in the American nuclear war program. He led the team of scientists who demonstrated the feasibility of the uranium chain fission process in the first artificial nuclear reactor (Figure 6). Otto Hahn was officially recognized to have played a major role in the discovery of fission, and in 1945 he was awarded the 1944 Nobel Prize for chemistry.

THE CONSEQUENCES OF URANIUM FISSION: NUCLEAR POWER

As with the discovery of X-rays some 40 years earlier, the advent of fission caused considerable excitement in the scientific community. Within a few months after the publications of January 1939, it became clear that a nuclear chain reaction was possible: one neutron splits a uranium nucleus, liberating nuclear energy and two neutrons, which, in turn, can produce two new fission events with a release of more energy and four new neutrons, and so on.

World War II was already raging in Europe when, in the United States, three Hungarian-born American scientists, Leo Szilard (1898—1964), Eugene Wigner (born 1902), and Edward Teller (born 1908), persuaded Albert Einstein that every effort had to be made to develop a weapon based on the uranium chain fission process before it could be achieved by Nazi Germany. Einstein wrote a letter to President Roosevelt in August 1939. The United States entered the nuclear weapon race in the fall of that year,

FIGURE 6. Enrico Fermi (1901 — 1954).

when Roosevelt appointed an informal "Advisory Committee on Uranium".

The project was still in its infancy when, in October 1941, British nuclear scientists concluded that enough ^{235}U could be accumulated by separation from natural uranium for the construction of a weapon suitable for conveyance by an aircraft. In December 1941, Roosevelt authorized the organization of an enormous project for devising the atomic weapon. It was 1 day before the Japanese attacked Pearl Harbor and the United States entered the war.

This was the greatest and most highly secret program of World War II. It was referred to as the Manhattan Project because much of the early work was done in laboratories around New York City. This enterprise involved as many as 65,000 people enlisted from various universities and famous corporations such as Eastman, Dupont, and Union Carbide, and cost 2000 million dollars (at the time). It received top priority on material and brains and included the most illustrious persons of American science, together with brilliant scientists who had left Europe, many because of war and the racial policies of the German and Italian dictatorships. A dozen Nobel Prize winners or winners-to-be were engaged in the project.

Results produced in this research were "classified" and could be presented only in internal reports with limited and controlled distribution. Some were gradually published after the war in scientific and technical literature, in particular a decade later in the proceedings of the first International Meeting for the Peaceful Uses of Atomic Energy, which was organized in Geneva (Switzerland) in 1955 by the United Nations Organization.

The goal of the Manhattan project was clear, but the means of attaining it were less evident. It was obvious that the ^{235}U isotope had to be separated from ^{238}U which is 140 times more abundant. Since the isotopes have identical chemical properties, only physical methods based on their small mass difference (less than 1 percent) could be considered. Two industrial plants were constructed. The first was based on a technique involving a volatile uranium compound for which diffusion proceeds at slightly different rates for the two isotopes. The plant was built in a sparsely inhabited region at Oak Ridge in Tennessee and proved to be successful. The second procedure utilized an electromagnetic separation of the two isotopes, but full-scale development was never reached.

Another challenging approach to nuclear explosives relied on plutonium, a chemical element that does not exist in nature, and of which the ^{239}Pu isotope was a promising fissile material similar to ^{235}U. Kilogram amounts of plutonium could, in principle, be produced by massive irradiation of uranium with neutrons.

Realization of the plutonium project started with the construction of an assembly of uranium and graphite, in which high fluxes of neutrons were predicted in the course of a controlled chain fission reaction. The project was conceded under the code name "The Metallurgical Laboratory", and was under the leadership of A.H. Compton (1892—1962) and his leading associate E. Fermi.

It was on a squash court under the stands of the University of Chicago's football field that Fermi built the first "nuclear pile". This was an edifice of graphite bricks (to slow down the neutrons) in which pieces of uranium metal and uranium oxide were stocked. Fermi started the first uranium chain fission process ever set up by man on December 2, 1942. This nuclear pile was soon transferred to a specially constructed building at Argonne, a site close to Chicago.

Although plutonium slowly accumulates in uranium in the nuclear reactor, the large-scale production of this new element had already been decided in 1942. The understanding of plutonium chemistry was based on studies of minute amounts of a plutonium isotope (milligrams and less), which were produced by cyclotron, a research device used in nuclear physics. The first batch was produced in September 1942, and contained only 2.8 micrograms of plutonium. Late in 1942, the Dupont Company agreed to assume responsibility for the building and operation of a large plutonium plant to be operated at Hanford in the state of Washington. This was a direct jump from milligram amounts to the kilogram scale and represented a unique case in the history of science. The short time required for this achievement — less than 2 years — was another astonishing accomplishment.

Assembly of the fissile material into a weapon was realized in Los Alamos, New Mexico, under the direction of J.R. Oppenheimer (1904—1967). Several thousand people lived and worked in strict seclusion. The construction of the uranium bomb did not present major difficulties. To trigger the explosion, two pieces of enriched uranium had to be very rapidly joined. This was a problem of classical ballistics and was solved using a gun barrel arrangement in which one sub-critical piece could be fired at another subcritical piece of fissile metal. The scientists were so confident in success that no test firing was conducted before the bomb was dropped on Hiroshima on August 6, 1945.

The same technique could not be used for the plutonium bomb because of the existence of an isotope which is fissioned by cosmic neutrons. The solution adopted was that of a fast implosion of a plutonium core produced by the detonation of chemical explosives. The reliability of this technique was so uncertain that a test firing was conducted near Alamogordo, New Mexico on July 16, 1945. This famous "Trinity test" was the very first nuclear explosion. On August 9, 1945, a plutonium bomb destroyed Nagasaki.

COMPLETION AND EXTENSION OF THE PERIODIC TABLE OF THE ELEMENTS

Within a few years after the discovery of radioactivity, six chemical elements had been added to Mendeleev's periodic system; five of these were radioactive: polonium and radium in 1898, radon and actinium in 1899, and protactinium in 1913. The sixth was hafnium, which is not radioactive, although it was also discovered in 1923 by a radiochemist, George Hevesy (1885—1966).

At the time when artificial radioactivity was discovered, only four chemical elements were missing in the periodic system. There was little doubt that the empty spaces near the end of the table should be occupied by radioactive elements, since the heaviest known stable element was bismuth. Indeed, the element located between radon and radium was found to be a highly unstable decay product of actinium: francium. Discovered in 1939 by Marguerite Perey (1909–1975), an assistant of Marie Curie, it was the last element to be found in nature.

A missing element towards the middle of the periodic table below manganese had puzzled chemists for a long time. It is a radioactive one, discovered in 1937 by Carlo Perrier (1886—1958) and Enrico Gino Segré (born 1905), in a molybdenum target which had been exposed to deuterons from the cyclotron in Berkeley. Being the first man-made chemical element, it bears the name technetium (from the Greek "artificial"). Astatine, located between polonium and radon, was the second synthetic element. It was discovered by Segré and his team (1940) in bismuth bombarded with α-particles. Finally, the periodic table was completed in 1946 when Ch. Coryell discovered promethium, a radioactive rare earth, among the fission products of uranium.

The chemists were highly satisfied with the completion of Mendeleev's periodic system less than 1 century after its appearance, but they were even more excited by the "terra incognita" beyond uranium. Today we know that Fermi was not entirely wrong in the mid-1930s when he claimed the discovery of transuranium elements, since the mixture of radioactive species in his uranium samples exposed to neutrons must certainly have contained some of them. Nevertheless, the first definitive proof of the synthesis of an element heavier than uranium was provided in 1940 by E.M. McMillan (born 1907) and P. Abelson (born 1913), who gave the name neptunium, symbol Np, to the new element.

The recognition of neptunium was a starting point for a new adventure in chemistry, which led to the discovery of plutonium in 1941 (but the publication appeared in 1946), and thence to 15 more transuranium elements. Mendeleev's table, in which the natural elements terminate at uranium with atomic number (Z) 92, was therewith extended (Table 1). The synthesis of element 108 was reported in 1984; a team of 12 physicists working under the direction of P. Armbruster (born 1931) in Darmstadt (West Germany) produced and identified three atoms of this element. Such discoveries illustrate the degree of technical skill that is possible in tracer-level manipulations. However, since they were first synthesized, the transuranium elements up to einsteinium have been produced in weighable quantities (>1 microgram) and present-day amounts of available plutonium can be expressed in tons.

Table I

PERIODIC CHART OF THE ELEMENTS (1988)

1	2	3	4	5	6	7	8	9	10	11	12	13	14	15	16	17	18
1 1.0 **H** HYDROGEN																	2 4.0 **He** HELIUM
3 6.94 **Li** LITHIUM	4 9.0 **Be** BERYLLIUM											5 10.8 **B** BORON	6 12.0 **C** CARBON	7 14.0 **N** NITROGEN	8 15.99 **O** OXYGEN	9 18.9 **F** FLUORINE	10 20.18 **Ne** NEON
11 22.99 **Na** SODIUM	12 24.3 **Mg** MAGNESIUM											13 26.98 **Al** ALUMINIUM	14 28.08 **Si** SILICON	15 30.97 **P** PHOSPHORUS	16 32.06 **S** SULFUR	17 35.45 **Cl** CHLORINE	18 39.95 **Ar** ARGON
19 39.1 **K** POTASSIUM	20 40.08 **Ca** CALCIUM	21 44.95 **Sc** SCANDIUM	22 47.88 **Ti** TITANIUM	23 50.94 **V** VANADIUM	24 51.99 **Cr** CHROMIUM	25 54.94 **Mn** MANGANESE	26 55.85 **Fe** IRON	27 58.93 **Co** COBALT	28 58.7 **Ni** NICKEL	29 63.54 **Cu** COPPER	30 65.4 **Zn** ZINC	31 69.72 **Ga** GALLIUM	32 72.6 **Ge** GERMANIUM	33 74.92 **As** ARSENIC	34 78.96 **Se** SELENIUM	35 79.9 **Br** BROMINE	36 83.8 **Kr** KRYPTON
37 85.47 **Rb** RUBIDIUM	38 87.62 **Sr** STRONTIUM	39 88.9 **Y** YTTRIUM	40 91.22 **Zr** ZIRCONIUM	41 92.9 **Nb** NIOBIUM	42 95.94 **Mo** MOLYBDENUM	43 (98) **Tc** TECHNETIUM	44 101.07 **Ru** RUTHENIUM	45 102.9 **Rh** RHODIUM	46 106.42 **Pd** PALLADIUM	47 107.87 **Ag** SILVER	48 112.4 **Cd** CADMIUM	49 114.8 **In** INDIUM	50 118.7 **Sn** TIN	51 121.7 **Sb** ANTIMONY	52 127.6 **Te** TELLURIUM	53 126.9 **I** IODINE	54 131.3 **Xe** XENON
55 132.9 **Cs** CAESIUM	56 137.33 **Ba** BARIUM	57–71 **La–Lu** LANTHANIDES	72 178.5 **Hf** HAFNIUM	73 180.95 **Ta** TANTALUM	74 183.85 **W** TUNGSTEN	75 186.2 **Re** RHENIUM	76 190.2 **Os** OSMIUM	77 192.2 **Ir** IRIDIUM	78 195.1 **Pt** PLATINUM	79 196.96 **Au** GOLD	80 200.6 **Hg** MERCURY	81 204.38 **Tl** THALLIUM	82 207.2 **Pb** LEAD	83 208.98 **Bi** BISMUTH	84 (209) **Po** POLONIUM	85 (210) **At** ASTATINE	86 (222) **Rn** RADON
87 (223) **Fr** FRANCIUM	88 (226) **Ra** RADIUM	89–103 **Ac–Lr** ACTINIDES	104 (261) **Unq** UNNILQUADIUM	105 (262) **Unp** UNNILPENTIUM	106 (263) **Unh** UNNILHEXIUM	107 (262) **Uns** UNNILSEPTIUM	108 (265) **Uno** UNNILOCTIUM	109 (266) **Une** UNNILENNIUM	110 (272) **Uun** UNUNNILIUM	111 (272) **Uuu** UNUNUNIUM	112 **Uub** UNUNBIUM	113 **Uut** UNUNTRIUM	114 **Uuq** UNUNQUADIUM	115 **Uup** UNUNPENTIUM	116 **Uuh** UNUNHEXIUM	117 **Uus** UNUNSEPTIUM	118 **Uuo** UNUNOCTIUM

Lanthanides:

57 138.9 **La** LANTHANUM	58 140.11 **Ce** CERIUM	59 140.91 **Pr** PRASEODYMIUM	60 144.24 **Nd** NEODYMIUM	61 (145) **Pm** PROMETHIUM	62 150.36 **Sm** SAMARIUM	63 151.96 **Eu** EUROPIUM	64 157.25 **Gd** GADOLINIUM	65 158.92 **Tb** TERBIUM	66 162.5 **Dy** DYSPROSIUM	67 164.93 **Ho** HOLMIUM	68 167.26 **Er** ERBIUM	69 168.93 **Tm** THULIUM	70 173.04 **Yb** YTTERBIUM	71 174.97 **Lu** LUTETIUM

Actinides:

89 (227) **Ac** ACTINIUM	90 232.04 **Th** THORIUM	91 231.03 **Pa** PROTACTINIUM	92 238.03 **U** URANIUM	93 (237) **Np** NEPTUNIUM	94 (244) **Pu** PLUTONIUM	95 (243) **Am** AMERICIUM	96 (247) **Cm** CURIUM	97 (247) **Bk** BERKELIUM	98 (251) **Cf** CALIFORNIUM	99 (252) **Es** EINSTEINIUM	100 (257) **Fm** FERMIUM	101 (258) **Md** MENDELEVIUM	102 (259) **No** NOBELIUM	103 (262) **Lr** LAWRENCIUM

Note: The figure in the left corner of the box is the atomic number and that in the right corner is the atomic mass. For radioelements lacking long-lived isotopes, the atomic mass given in parentheses is that of the most common or most stable isotope. The lanthanide and actinide elements are grouped at the bottom of the table. The official nomenclature is used for elements beyond atomic number 103.

At the beginning of 1990s the official nomenclature for elements beyond atomic number 103 was as shown on Table I. The histories of the discoveries of the elements were still not settled at that time. Element 104 has two unofficial names, rutherfordium and kurchatovium, and two official discoverers at Berkeley and Dubna (1969). The same is true of element 105, called hahnium by one group and nielsbohrium by the other. Element 106 has been waiting for a name since 1974 when it was reported by researchers at Berkeley-Livermore. Element 107 is anonymous since 1976, when discovered in the German Laboratory for Heavy Ion Research in Darmstadt. A joint credit to Dubna and Darmstadt was assigned for element 108 in 1984, while element 109 is another Darmstadt discovery (1982). For elements 110 to 112 the evidence was still doubtful.

OMNIPRESENCE OF RADIOACTIVITY AND SOME IMPLICATIONS

Within a century knowledge of radioactivity and radiations, together with a cognizance of their numerous important implications, had been extended from the first recognition in uranium and thorium salts to establishment of their omnipresence on the Earth and in the Cosmos.

As early as 1902, the Curies considered that each atom of a radioactive substance acts as a source of energy and the following year they noticed that a salt of radium continuously released enough heat to melt ice.

Radioactive elements are dispersed throughout the matter of the solar system and the "radiogenic heat" was later found to be of importance for the evolution of planetary bodies including the Earth.

The Earth receives 8000 calories per square meter annually from radioactive elements dispersed in a surface layer of 1 kilometer thickness. In the past, the generation of heat was up to three and a half times higher owing to contribution from short-lived radioelements.

Very large amounts of radiogenic heat were released at sites where natural nuclear reactors were operating, such as in the 2000 million-year-old fossil reactors in the Oklo uranium mine (Gabon, Africa). The Oklo phenomenon was discovered in 1972, after H. Bouzigues and his colleagues (in a nuclear-fuel processing plant in France) analyzed the samples of some uranium ores from the Oklo mine and found that, as with spent fuel elements in artificial nuclear reactors, the uranium was depleted in fissile isotope (^{235}U). In addition to radiogenic heat, these reactors were localized sources of ionizing radiation which could have acted as specific and effective energy sources for Precambrian chemical processes.

Radioactivity is useful in dating geological events and a first attempt to measure the absolute age of minerals by using a radioactive parent-daughter pair was made in 1906 by Rutherford. The latter realized the significance of the relationship between uranium or thorium, and helium accumulated as a result of the radioactive element's decay over a geological time scale. The following year, B.B. Boltwood published a list of geological ages based on the ratio of lead to uranium in some minerals. Although inaccurate, these determinations showed that radioactive dating is feasible and they marked the beginnings of radiogeochronology. From the omnipresence of radionuclides in nature and the wealth of reliable knowledge obtained from advances in radioactive dating, it is now possible to ascertain the dates of important events in the life of our planet, the solar system, and even the universe.

The first artificial transmutation of a chemical element in 1919, together with the start of spectroscopic analyses of the elemental composition of stars during the period 1925—1928, gave birth to the idea that all elements result from nuclear reactions. In turn, concepts of nucleogenesis prompted the development of theories on the origin and evolution of the universe. G. Gamow (1904—1968), first in 1946 and later in 1948 with R.A. Alphen and H.A. Bethe, proposed the theory of an expanding universe born in a gigantic explosion, the "Big Bang". This theory was untenable in its original form, but its basic concept is now universally accepted. According to the Big Bang model, radiations were preponderant at the earliest stage in the formation of the universe (within the first fractions of a second). The fairly recent discovery of a low-energy radiation which fills cosmic space is interpreted as a remnant of the Big Bang. This "fossil" radiation, which strongly supports Gamow's idea, was discovered in 1965 by A.A. Penzias and R.W. Wilson.

Not long after the discovery of radioactivity, it was found that the instruments used for its measurements indicated the presence of radiation even in the absence of radioactive sources. Shielding the detectors with thick lead absorbers reduced, but never eliminated, the effect. When the instruments were raised in balloons in the course of experiments conducted during the years 1911 to 1913, it was shown that the intensity of this radiation increases more than tenfold at an altitude of 9000 meters. These experiments led V.F. Hess to suggest that this penetrating radiation is of extraterrestrial origin. This was confirmed in 1925 by R.A. Millikan (1868—1950), who pursued the measurements up to 15,000 meters and coined the phenomenon "cosmic radiation". Since then, studies of cosmic rays provide a wealth of information on events far beyond the Earth, in interstellar and intergalactic space. The role of satellites is becoming increasingly important in procuring information on cosmic rays. Part of the data is relevant to the history of the universe, and is even helpful in retracing its origin in the Big Bang.

Celestial objects such as meteorites and comets are continuously exposed to cosmic radiation. This results in various degrees of radiation-induced damage, chemical changes, and radioactive nuclides produced by nuclear reactions. Some of these radionuclides are useful as radioactive clocks.

Cosmogenic nuclides are also formed by reactions of cosmic rays with atmospheric constituents. One of these nuclides is radioactive carbon. It was discovered by W.F. Libby (1908–1980) in 1947 and has since become an essential tool in archeological dating.

USES AND MISUSES OF RADIOELEMENTS AND RADIATIONS

A peculiar property of radioelements is their ease of detection, even in imponderable amounts. As a result of this quality, radioactivity has rapidly ceased to be a field of art for art's sake and the development of the uses of radioelements has thus kept stride with advances in basic knowledge.

George Hevesy realized as early as 1913 that a radioisotope could be used as an indicator (tracer) of the corresponding chemical element and he performed the first experiment (with Frederick Adolphus Paneth [1887—1958]) on the determination of the solubility of lead salts using radioactive lead. In the following year, he introduced the tracer technique in biology to study the uptake of lead by beans, and the distribution of bismuth in a rabbit.

The application of radioisotopes as tracers began to flourish in the mid-1930s after the discovery of artificial radioactivity and the invention of the cyclotron for isotope productions. Hevesy introduced the use of an artificial radioisotope (radiophosphorus) to follow the metabolism of phosphorus in rats and in humans, thus opening the way to nuclear medicine. He was also a pioneer in another important field of application of radioactivity, viz., activation analysis (1936). This technique enables reliable detection or determination of trace amounts of elements after they have been made radioactive by suitable nuclear reactions. For all these achievements, Hevesy was awarded the Nobel Prize in 1943.

Within 5 years, several hundred articles were published on tracer investigations with radioactive bromine, chlorine, iodine, carbon, cobalt, iron, and sodium.

Radioactive sodium was used in the treatment of a leukemia patient for the first time in 1936 and several other artificial radionuclides were rapidly found to be superior to radium, since they could be used in chemical forms that are more suitable for medical applications.

Tracers were also applied very early in industry. In 1941, radioactive phosphorus was used to label piston rings for studies of lubrication and wear.

The earliest observation of a chemical change induced by ionizing radiation is as old as the discovery of ionizing radiation itself. It was made in 1895 when it was found that X-rays are able to fog photographic plates, i.e., induce a chemical change in the photographic emulsion.

Chemists working with aqueous solutions of radioelements noticed peculiar changes which could be attributed to the chemical action of radiation, and the early days of radioactivity and radiochemistry were also those of radiation chemistry. F. Giesel (1852—1927) already reported in 1903 that a gaseous mixture of hydrogen and oxygen is constantly liberated from aqueous solutions of radium bromide, and that the solution becomes brownish because of the generation of bromine. In 1914, A. Debierne (1874—1949) proposed an explanation for the phenomena observed in aqueous solution: the passage of radiation through water produces ionized molecules which react and form oxidizing and reducing entities, i.e., OH and H. The latter are chemically very reactive species which attack the molecules present in an aqueous system and are responsible for chemical changes. The hypothesis was far ahead of its time. Some of the proposed reactions proved to be untenable but the basic idea was confirmed in the early 1940s. In the following 2 decades, the radiation chemistry of water was very thoroughly investigated not only since it is a main constituent of the biosphere, but because of its important role in reactor technology as a coolant and moderator and in nuclear processing as a solvent.

Many tragic misuses of radiation already began with the discovery of X-rays owing to a popular belief in their "magic" power. One widespread use, which continued well into the 1930s, consisted in a treatment for removing superfluous hair. Thousands of young women were exposed to X-ray doses sufficient to produce depilation, which inevitably also caused side effects such as erythemas, scarring, and cancers. The general public attributed even more prodigious properties to radium. The noble and legendary figure of Marie Curie probably contributed to that belief. Radioactive liquors were offered for the treatment of various illnesses, some of which were highly radioactive. Hair tonics, tissue creams, and even chocolate bars and toothpastes containing radium (or supposedly so) were sold up into the 1940s. Uncontrolled use of radium for self-luminous paints in industry, and for dial painting continued even longer. Lethal effects occurred in workers (mainly young women) who moistened their brush tips with their lips, thereby ingesting considerable amounts of radium.

The first report of possible injury from X-rays appeared less than 3 months after Roentgen's discovery, and described eye irritations associated with the use of X-rays and fluorescent materials. Shortly afterward accounts of skin burns began to appear. In 1901, Becquerel inadvertently carried a tube of radioactive material in a waistcoat pocket until he noticed the reddening of the underlying area of skin. To be sure that it was due to radiation, he repeated the practice using tubes with and without radioactive material. Pierre Curie deliberately kept a sample of radium in contact with his arm until radionecrosis and ulceration of the skin appeared. Marie Curie's fingers and hands were severely burned and she had periodic spells of fatigue and weakness before she eventually died of a malignant blood disease. Biographers have reported necrosis of the fingers and hands among several prominent scientists of the first generation of radiochemists. More than 300 early radiation workers died from overexposure to radiation. Many pioneers, however, lived up to a high age, showing that individuals exhibit different degrees of sensitivity to radiation.

The field of radiation biology has developed from such observations and from laboratory studies. Knowledge has also progressed with experience gained from unfortunate events which have affected large populations, like the atomic bombing of Hiroshima and Nagasaki (1945) or the catastrophe of the Chernobyl nuclear power plant (1986).

The most detailed study concerns victims of the holocaust in Japan. The "Atomic Bomb Casualty Commission" has controlled the survivors and their progeny during the last decades. The mortality survey has been conducted on about 48,000 people who were exposed to significant amounts of radiation from nuclear bombs, in addition to 35,000 inhabitants who suffered smaller exposures, and a further 25,000 who subsequently came to the cities to help in rescue work or in the search of relatives.

The nuclear reactor accident in the Chernobyl power plant, which has released large amounts of radioactive material to the atmosphere and the surrounding environment, will also provide valuable information on individual human reactions following irradiation. Immediately after the accident in 1986, a systematic health control and medical aid plan was set up involving about 116,000 people. Studies include assessments of the health situations of people still living in the affected areas, human radiation exposure, and the environmental contamination.

RADIATION PROTECTION REGULATIONS

Despite the fact that harmful effects of radiations were recognized soon after their discovery, radiation protection measures and subsequent national and international regulations were slow to develop. Many obstacles had to be overcome, such as a suitable definition of an amount of radiation (dose), its biological significance, the elaboration of appropriate instrumentation for reliable measurement and the procedures for its calibration.

Paul Villard (1860—1934) already suggested in 1908 that X-rays can be quantified by the electric charge which they produce in air under defined conditions. This idea was adopted on an international level only 20 years later and was extended to the radiations of radioelements after a further delay of 10 years.

The first steps toward regulation were taken in Germany (1913) and Britain (1915), primarily to protect the patients and the medical staff. National advisory committees on radiation protection were formed and these issued the first recommendations. By the early 1920s, several countries had followed the same policy.

Important events for international regulations in radiation protection were the International Congresses of Radiology in London (1925) and in Stockholm (1928). Two working bodies emerged from these meetings: a standing International Commission on Radiation Units and Measurements (ICRU) and the International Commission on Radiological Protection (ICRP). Their aim was to keep the definition of units and the methods of measurement up to date, as well as to provide recommendations for protection. The bodies are nongovernmental and their members, eminent scientists, are faced with the difficult task of assessing the rapid extension of ways in which people become exposed to ionizing radiation, and keeping pace with the rapid accumulation of knowledge of the effects of radiation from various radiobiological, genetic, and medical studies. Both commissions are indispensable in formulating and improving the national regulations relevant to radioactive materials and radiation. The transfer of their recommendations to national legislatures is supported by governmental organizations which operate within the frame of the United Nations Organization. One of these institutions is the International Atomic Energy Agency which was founded in 1957 to promote the peaceful uses of nuclear energy.

NUCLEAR POWER FOR ELECTRICITY

The first nuclear power plant for generating electricity was a 5 megawatt-electric installation which went into operation in June 1954 at Obninsk near Moscow (Soviet Union). By October 1956, Great Britain was operating its Calder Hall plant for plutonium production, whereby electricity was generated as a by-product at a power output of 50 megawatts of electricity. The first nuclear reactor for civilian generation of electricity in the United States produced 60 megawatts of electricity and was put into operation in Shippingport (Pennsylvania) in May 1958. By the end of the 1950s a commercial nuclear power plant appeared in France and several new ones in the Soviet Union and United States, so that some 15 nuclear reactors were globally producing less than 1000 megawatts of electricity, which represented about 0.1 percent of the world total production of electricity.

Only 15 years later, in 1975, 185 nuclear reactors were producing electricity with a total capacity of 75,000 megawatts. The nuclear share in global production of electricity had increased to about 6 percent. At the end of 1991 it was about 12% with 327,213 megawatts of electricity installed and 420 nuclear power plants connected to electricity grids in 25 countries. Although impressive, this figure seems modest when compared to predictions for 1990 and 2000 in the Annual Reports of the International Atomic Energy Agency, with estimates of 1,600,000 and 3,600,000 to 5,000,000 megawatts of electricity, respectively.

On a general basis, and especially in the aftermath of the Chernobyl accident, the prospects for civil nuclear power have become somber in many countries. The most gloomy outlook prevails in the United States, which leads the world in many fields of nuclear power and where the nuclear share at the end of 1991 was 21.7% of the total electric capacity. Quite a few plants which have been under construction within the last decade cannot get on-line. The most salient question seems to concern plant safety, with its impact on public confidence and financial credibility.

Another specific application of nuclear power is its use for military purposes. For example, small nuclear reactors have been occasionally used since 1965 to provide power for board instruments in American and Soviet spacecrafts.

Power production for propulsion is much more important. The first nuclear-powered vessel was Nautilus, a submarine launched in the United States in 1954 and famous for its underwater crossing of the Arctic Ocean in 1958. The first nuclear-powered surface vessel was the icebreaker Lenin, which was launched in 1957 in the Soviet Union. A merchant ship with nuclear propulsion was the American vessel Savannah, which was put out of service in 1967 after only 2 years, owing to exorbitant running costs. At present, nuclear propulsion seems suitable only for military purposes: some 575 submarines and surface vessels (cruisers and aircraft carriers) were crossing the oceans at the beginning of the 1990s.

NUCLEAR WEAPONS

After the explosions of American nuclear weapons in 1945, the "nuclear fission club" was successively joined by the Soviet Union (1949), Great Britain (1952), France (1960), and China (1964). In 1974, India performed a nuclear explosion for peaceful uses, according to the Indian government.

The development of nuclear weapons soon led to the appearance of another "club", the "thermonuclear" one. In 1952, the United States performed the first thermonuclear explosion on our planet, on a coral atoll in the Pacific Ocean. The explosion of a device weighing 65 tons yielded the equivalent of 500 Hiroshima bombs and the blast wiped the atoll off the map. The following year, the Soviet Union exploded a power thermonuclear bomb that was sufficiently small in size to be carried in an aircraft. Since 1954, thermonuclear bombs have become standard items in the armament stocks of the United States, the Soviet Union, and Great Britain, later followed by China (1967) and France.

During 13 years of nuclear testing (1945—1958), about 150 bombs of all varieties were brought to explosion, releasing an estimated 25 million curies of ^{90}Sr and ^{137}Cs into the atmosphere. A gentleman's agreement to suspend nuclear bomb testing was reached in 1958 among the United States, the Soviet Union, and Great Britain. Nonetheless, it did not prevent France from exploding its first nuclear device in the atmosphere in 1960, nor the Soviet Union from terminating the moratorium in 1961. The United States too, has continued with the tests. However, under the pressure of world public opinion, the three chief nuclear powers signed in 1963 a partial test-ban treaty, in which nuclear explosions were banned in space, the atmosphere, and underwater. Only underground explosions of a limited power were permitted, since these are not expected to produce fallout.

Limitations imposed on testing of the weapons did not prevent their development and production, and in the mid-1980s the world nuclear arsenal comprised the equivalent of about 15,000 million tons of TNT or 1 million Hiroshima bombs.

The consequences of using nuclear weapons are well understood from various systematic studies on military polygons of effects of blast, heat, radiation, and fallout. For years only estimates of devastation and deaths in the case of a nuclear conflict have been taken into account. In 1983, American scientists, including the astronomer Carl Sagan and the biologist Paul Erlich, pointed out the global, long term, climatic, and biologic consequences of multiple nuclear explosions, and drew attention to the "nuclear winter". The implications are clear: in the event of nuclear war there would be no winner and neutral nations would also suffer. More recent attempts to minimize the risks are based on modifications in computer modeling and in the selection of input data. Arguments are advanced for a nuclear "autumn" rather than a nuclear winter, particularly if certain rules are respected in conducting the nuclear war. The prevailing opinion, however, is that no one can predict the course of a nuclear war once it has begun, and there is no doubt that the likely long-term consequences can be only global and catastrophic (Chapter 10).

It should be noted that at the beginning of 1990s we are witnessing a drastic change in the behavior of the two nuclear superpowers, which have moved from conflict to cooperation. It started in 1987 with the signing of a treaty banning intermediate-range nuclear missiles and was completed in 1992 with agreement to dispose of all multiwarhead missiles on their territories by the year 2003. But, as the fear of a nuclear war between the superpowers recedes, serious concerns remain. The politically and socially disintegrated society of the former Soviet Union still owns practically half of the global nuclear arsenal and the capacity for production. The risks accumulated over four decades of the Cold War are numerous and dealing with this legacy of the nuclear age will most likely take a generation.

Further Reading

A wealth of books and reviews has been published on the historical development of nuclear sciences, dating from the discovery of radioactivity to the elaboration of the atomic bomb during World War II, with all its postwar geopolitical implications. The following selection is intended to satisfy the interest of a large scope of readers.

Badash, L., Ed., *Rutherford and Boltwood, Letters on Radioactivity*, Yale University Press, New Haven, Connecticut, 1960. [The correspondence between the two great scientists that reveals not only their personal accomplishments, but also the era of early radioactivity to which they both contributed.]

Badash, L., Ed., *Radioactivity in America, Growth and Decay of a Science*, John Hopkins University Press, Baltimore, 1979. [A well-documented and easily readable book on the American contribution to radioactivity.]

Badash, L., Hirschfelder, Y.O., and Broida, H.B., *Reminiscences of Los Alamos (1943—1945)*, D. Reidel Publishing, Dodrecht, 1980. [An anecdotic report on the life of the scientists secluded in Los Alamos while making the atomic bomb.]

Cotton E., *Les Curies*, Seghers, Paris, 1963. [Emotional accounts of the life, struggle, and achievements of Pierre and Marie Curie.]

Curie E., *Madame Curie*, Gallimard, Paris, 1938. [A sensible daughter presents her mother and the life in the family. A classic.]

Curie M., *Pierre Curie*, Payot, Paris, 1924. [A women tells on the work of her husband, collaborator, and great scientist.]

Fermi, L., *Atoms in the Family: My Life with Enrico Fermi*, University of Chicago Press, Chicago, 1954. [A personal and emotional account of the leading figure in the second generation of nuclear scientists.]

Giroud, F., *Une Femme Honorable*, Fayard, Paris, 1981. [A biography of Marie Curie as woman, well documented, and emotionally written.]

Goldschmidt, B., *Les Rivalites Atomiques (1939—1966)*, Fayard, Paris, 1967. [An active participant presents a well-documented and clearly written historical survey.]

Hewlett, G., and Anderson, O.E., *The New World 1939—1946*, Vol. 1 and 2, Pennsylvania State University Press, University Park, 1969. [A history of the U.S. Atomic Energy Commission.]

Ivimey, A., *Marie Curie, Pioneer of the Atomic Age*, Praeger Publishers, New York, 1980. [The story of Marie Curie, in a clear and lively style.]

MacKay, A., *The Making of the Atomic Age*, Oxford University Press, Oxford, 1984. [A well-written, concise book which deals with various biographical, historical, technical, and political aspects.]

Reid, R., *Marie Curie*, E.P. Dutton, New York, 1974; *Marie Curie derrière la Légende*, Seuil, Paris, 1974 (French translation). [A provocative and documented love story of Marie Curie.]

Romer, A., *The Discovery of Radioactivity and Transmutation*, and *Radiochemistry and the Discovery of Isotopes*, Dover Publication, New York, 1964. [An original mixture of reprinted papers and comprehensive comments which illustrates the development of radioactivity and gives a full bibliography of major contributions from 1886—1913.]

Segré E., *From X-Rays to Quarks, Modern Physicists and Their Discoveries*, A. Mondadori, 1980. [A contemporary physicist's account on the evolution of physics in the 20th century.]

Trenn, Th.J., Ed. *Radioactivity and Atomic Theory*, Taylor & Francis, London, 1975. [Soddy's annual progress reports on radioactivity from 1904—1920 to the Chemical Society. A brillant summary of the achievements.]

Weart S., *Scientists in Power*, Harvard University Press, Cambridge, Massachusetts, 1979. [An outstanding, extraordinarily well-documented book focusing on the French contribution to the nuclear sciences from Marie Curie to Frédéric Joliot, with an insight into the French scientific and political society.]

Annex

The following are Nobel Prize awards for discoveries related to radioactivity. (The asterisks denote prizes in physics; otherwise, prizes in chemistry.)

1901(*)   **W.C. Roentgen,** Professor at the University of Munich, for his discovery of X-rays.

1903(*)   Shared between **H.A. Becquerel,** Professor at Ecole Polytechnique in Paris, for discovery of spontaneous radioactivity; and **P. Curie,** Professor at Ecole Municipale de Physique et Chimie Industrielle, and **M. Curie,** Paris, for their joint work on the phenomena of radiations discovered by **H.A. Becquerel**

1908   **E. Rutherford,** Professor at the University of Manchester, for his research on the disintegration of atoms and the chemistry of radioactive elements

1911   **M. Curie,** Professor at the University of Paris, for her contribution to chemistry by the discovery of radium and polonium, by the definition of radium and its isolation in the metallic state, and for research on the compounds of radium

1921   **F. Soddy,** Professor at the University of Oxford, for his contribution to the chemistry of radioactive compounds and his research on the existence and on the nature of isotopes

1935(*)   **J. Chadwick,** Professor at the University of Liverpool, for his discovery of the neutron

1935   Shared between **F. Joliot,** Professor at the University of Paris, and **I. Joliot-Curie,** Paris, for the synthesis of new radioactive elements

1938(*)   **E. Fermi,** Professor at the University of Rome, for the production of new elements by neutron irradiation and for his discovery of nuclear reactions induced by slow neutrons

1939(*)   **E.O. Lawrence,** Professor at the University of California, Berkeley, for the invention and development of the cyclotron and for it use, mainly in the production of artificial radioactive elements

1943	Awarded in 1944 to **G. de Hevesy,** Professor, Stockholm, for his work on the use of isotopes as tracers in the study of chemical reactions
1944	Awarded in 1945 to **O. Hahn,** Professor, Berlin-Dahlem, for his discovery of the fission of heavy atomic nuclei
1951	Shared between **E.M. MacMillan** and **G.T. Seaborg,** Professors at the University of California, Berkeley, for their discoveries in the field of the chemistry of the trans-uranium elements
1960	**W.F. Libby,** Professor at the University of California, Los Angeles, for the utilization of ^{14}C in age determination in archeology, geology, geophysics, and other sciences
1967(*)	**H.A. Bethe,** Professor at Cornell University, Ithaca, for his contributions to the theory of nuclear reactions, especially with respect to the origin of stellar energy

A clear knowledge of the structure of atoms has been acquired since the first decade of the present century. Experiments by Rutherford implied that the atom was composed of a positive nucleus and orbiting electrons. A theory by Niels Bohr explained the electronic behavior. Later developments in quantum theory replaced notions of fixed electron orbits by a more insubstantial concept, and the advent of wave mechanics enabled the interpretation of atomic phenomena of greater complexity. In this chapter we consider only the consequences of theoretical approaches to atomic and subatomic structures.

The "solar system" model is still used to visualize the atom. In this representation, light, negatively charged electrons "orbit" around a heavy, positive nucleus. The nucleus contains two kinds of nucleons, one of which is positively charged (the proton) and the other neutral (the neutron). However, experiments show that, in addition to electrons and nucleons, the subatomic world consists of hundreds of constituents, some of them being composite entities. Particle physics suggests that this multitude can be reduced to only two types of elementary particles: quarks and leptons.

Four fundamental forces operate in nature and are classified as electromagnetic force, strong force, weak force, and gravitational force. They are thought to operate by way of force-carrier particles called gauge bosons.

The gauge bosons, quarks, and leptons are believed to be the truly elementary particles, i.e., the constituents of matter which do not consist of anything smaller.

About 300 kinds of nuclei are stable. Hundreds of other natural or artificial ones are ephemeral because of radioactivity. Radioactive decay is a spontaneous disintegration of the nuclei with emission of energy in the form of particles or electromagnetic radiation. It is a statistical phenomenon and the time required for half of a given number of atoms to be transformed is called the half life. It may range from millions of years to a fraction of a second.

Presently existing radioactive elements and the remnants of decayed radionuclides are used as "radioactive clocks" in dating various objects and events not only on a historical time scale, but also within the solar system and in outer space.

2

Radioactivity and Matter

A PRIMER OF RADIOACTIVITY

Radioactivity is a phenomenon in which energy is released by certain substances in the form of invisible radiation. It is a natural process, omnipresent in the universe, which already existed when the chemical elements began to form.

The meaning of radiation is broad: it includes waves such as light and X-rays, as well as charged particles. The human senses are incapable of perceiving radioactivity. Detection of a radioactive species can only be achieved indirectly, by the chemical, physical, or biological changes induced by the radiation. The blackening of a photographic plate, the coloration of kitchen salt, the decomposition of water, or the killing of biological cells are visible consequences of the radiations emitted by radioactive substances. In gases, or in some solids (especially semiconductors), radiation produces changes which are converted into electric pulses which can be processed. By such means, images can be recorded on a screen to provide full information on the nature and on the energy of the radiation.

The energy released in a radioactive transformation must satisfy the law of energy conservation which states that in an isolated system of constant mass, energy can be neither created nor released. For the Curies and their contemporaries, the phenomenon of radioactivity seemed to be in contradiction with this principle. The energy emitted by radium showed no signs of weakening and it seemed that the radioelement could release energy perpetually. This misconception is now understandable because we know that the half-life of radium is so long (16 centuries) that observation during a human life span could scarcely notice a decrease in the mass of the radioelement.

Radioactive elements can be identified from the properties of their radiations. When radiation is emitted, the radioactive element simultaneously changes into a new element which is itself often radioactive. In a form of terminology that seems to overlook the principle of egality of the sexes, the two elements are referred to as "parent" and "daughter".

The term "radioactive decay" is commonly used when dealing with radioactivity. It signifies both the phenomenon in itself involving disintegration of an unstable nucleus and "transmutation". The latter term, which actually means the conversion of one element into another, was already used by the alchemists in their efforts to change common metals into gold and silver. Radioactivity is a spontaneous transmutation which proceeds immutably without human intervention. However, the change is not instantaneous. Some radioelements decay rapidly, whereas others transform slowly. Uranium is an example of the latter kind; it has not yet completely decayed since the Earth was formed thousands of millions of years ago. Other radioelements disappear completely in the course of several years, or within days or periods as short as only a fraction of a second.

In order to express the decay rate in a practical manner it is customary to use a term called the half-life of the radioelement. This is the length of time required for half of the radioactive element to decay. Radium discovered by the Curies has a half-life of 1600 years. This means that after 1600 years one half of the initial number of radium atoms have decayed; after 3200 years one half of the remainder or three fourths of the total have vanished, and so on. After seven half-lives have elapsed, 1 percent of the radioelement remains and after ten half-lives only one thousandth of the original quantity. Thus, the first source of radium prepared by Marie Curie in 1898 will have diminished to one half of its initial amount in the year 3498. A small quantity will still be present in the year 17,898.

A quantitative expression of the intensity of a radioactive element is the activity, which represents the number of decays occurring per second. The unit of activity is the becquerel, in recognition of the discoverer of radioactivity. One becquerel (symbol Bq) represents one disintegration per second. It is a very small activity and usually multiples of it must be used, for example, megabecquerel (MBq) for 10^6 becquerels, gigabecquerel (GBq) for 10^9 becquerels, and terabecquerel (TBq) for 10^{12} becquerels. In the past, and occasionally even at present for large amounts of radioactive substances, the curie (symbol Ci) was the reference unit of activity; 1 curie is equivalent to 37 gigabecquerels and fairly close to the activity of 1 gram of radium. Millions of curies are produced in nuclear power reactors and are present in radioactive wastes. Ambient radioactivity, whether natural or artificial, is generally very weak and is preferably expressed in becquerel units.

Although radioactivity can be examined with macroscopic tools, its essence lies in the intimate and ultimate structure of matter. A closer look into the atomic and nuclear world will help to elucidate the nature of the phenomenon of radioactivity.

ATOMIC STRUCTURE

The word "atom" became familiar to the layman when the first "atomic" bomb exploded in 1945. This event also marked the beginning of a new era in the history of mankind: the atomic age. For the philosophers and scientists, however, the hypothesis of an atomic structure of matter had already been progressively accepted long ago.

The basic concept of atoms was introduced by Greek philosophers over 2500 years ago. The word literally means "indivisible" and suggests that matter cannot be dissected indefinitely: slicing a stone or a piece of wood or an animal into smaller and smaller parts ends with an ultimate cut, leaving an object which cannot be further divided — the atom. The Greek atomists believed that all matter, living or inert, was composed of atoms and that the myriad forms of matter reflected the various configurations and movements of the atoms.

The discoveries of the electron and proton, the observation of radioactive decay, and eventually the discovery of the neutron progressively disclosed the microscopic structure of matter.

The Atom

Like any material object, atoms have a size and weight, although the numerical values of both properties are not easily perceptible. One drop of water, say five hundredths of a gram, contains 2×10^{21} oxygen atoms and twice as many hydrogen atoms. Accordingly, the size of an atom is quite minute: 10^{-10}

meter. It is vanishingly small on the human scale, but such dimensions are quite familiar to the experimentalist. It is a routine task in the laboratory to measure atomic distances, such as the length of a chemical bond or the distance between neighboring ions in a crystal lattice.

No balance is sensitive enough to weigh an atom because the latter's mass is generally less than 10^{-22} gram. It would require a pile of 6×10^{23} atoms to give 12 grams of carbon or 238 grams of uranium. This large number is the Avogadro constant, named after Amadeo Avogadro (1776—1856). The increase in the mass of the atom is not matched by a similar expansion of the atomic volume: the sizes of the hydrogen and the uranium atoms are about the same.

Inside the Atom

Atoms have a composite structure. The center consists of a heavy nucleus with a positive electric charge, which is surrounded by a swarm of much lighter particles, the negatively charged electrons. The electrons are attracted to the nucleus because the respective charges are of opposite sign, and travel with speeds approaching the velocity of light.

A popular picture of the atom is that of a miniature solar system, in which the planets (the electrons) revolve incessantly around the Sun (the nucleus) in well-defined trajectories. The atom is just as "empty" as the solar system, in which Mercury, the planet nearest to the Sun, is at a mean distance of 60 million kilometers. A crude representation of the hydrogen atom is a marble (the nucleus) at the center of a football stadium and a pinhead (the electron) orbiting over the benches.

Electrons revolve around the nucleus in orbits with discrete energies. According to quantum theory, an electron which gains an energy quantum can jump to an orbit further removed from the nucleus. It releases energy when it falls closer to the nucleus.

All atoms of a given element possess the same number of electrons, which is called the atomic number, and is given the symbol Z. The atomic number is directly pertinent to the periodic chart of the elements (Table 1, Chapter 1). Each box of the chart, corresponding to a single element, bears a number which is equal to the atomic number. The first element in the chart, hydrogen, is placed in box number one, and has a single electron (Z = 1). Uranium with 92 electrons is found in box number 92. Thus, from the position of the element in the chart one can deduce the number of electrons in the corresponding atom.

Inside the Nucleus

The radius of the nucleus is much smaller than that of the atom, being about 1/100,000th for the lighter nuclei and up to 1/10,000th for the heaviest. Almost all the mass of the atom is concentrated in the nucleus, whose density is about 10^{12} -fold higher than that of the atom as a whole.

The nucleus consists of nucleons which are of two types: protons, with a positive charge equal in magnitude to the negative charge of the electron, and neutrons, which have no electric charge. In a free atom, the number of electrons equals the number of protons and the atom is electrically neutral. Therefore, the atomic number Z also represents the number of protons in the nucleus (Figure 7).

With the exception of the nucleus of hydrogen, which consists of one single proton, all nuclei also contain neutrons. Both types

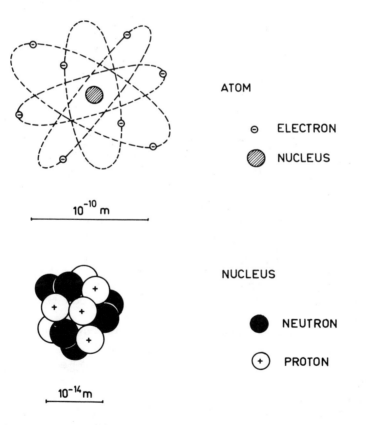

ATOM

⊖   ELECTRON

◍   NUCLEUS

10^{-10} m

NUCLEUS

●   NEUTRON

⊕   PROTON

10^{-14} m

FIGURE 7. Atom visualized as a solar system, and nucleons inside a nucleus.

of nucleons have nearly the same mass. The total number of nucleons inside the nucleus is called the mass number and is represented by the symbol A, which together with Z completely specifies the nucleus. The usual way of representing a nucleus is to start with the chemical symbol of the element, add the mass number at the upper left side, and add the atomic number at the lower left: ^1_1H for hydrogen, $^{197}_{79}Au$ for gold, $^{238}_{92}U$ for uranium. We frequently use the representation: gold-197 (^{197}Au) and uranium-238 (^{238}U).

Isotopes

In many cases, a nucleus with its fixed number of protons can accommodate different numbers of neutrons. This leads to species with the same atomic number but with different mass numbers. They pertain to the same element and all occupy the same box in the periodic chart of the elements. Such nuclei with the same number of protons, but different numbers of neutrons, are called isotopes. Accordingly, the isotopes of an element differ by their mass number. The nucleus of hydrogen, the proton, can combine with one neutron to form a nuclide with atomic number 1 and mass number 2, which is written ^2_1H. For an obvious reason this nuclide is termed "heavy" hydrogen, or, more frequently, deuterium. The deuterium atom, like the normal or light hydrogen, has a single electron. Terrestrial hydrogen is a mixture of both isotopes ^1_1H and ^2_1H, the first being the more abundant (over 99.8 percent). Tritium ^3_1H is a radioactive isotope of hydrogen, occurring to an extent of less than one part in 10^{17} in natural hydrogen. Tritium decays by beta (β)–emission with a half-life of 12.3 years (Figure 8).

With about 20 exceptions (e.g., aluminum and gold) most chemical elements in nature possess several isotopes. Copper is a mixture of $^{63}_{29}Cu$ (69.09 percent) and $^{64}_{29}Cu$ (30.91 percent). Natural tin is composed of 10 isotopes, all with 50 protons and from 62 to 74 neutrons. The total number of stable isotopes in the nature is about 300 and thus greatly exceeds the 81 stable elements.

If the chemical elements were synthesized from the same primordial material and at about the same time, their isotopic composition should be the same everywhere on Earth and also in the universe. Analyses of terrestrial and extraterrestrial samples indicate that this is basically the case, but with a few exceptions.

The various isotopes of an element, although belonging to the same chemical element, exhibit slightly different physical and chemical properties. As a result of physical-chemical processes which repeatedly occur in nature, small differences are amplified and variations in the isotopic composition are observed. The percentage of deuterium in hydrogen may vary from 0.0133 in fresh water (rain, spring) to 0.0156 in the ocean. Slight variations in the isotopic composition of elements from different samples have also been observed for carbon, oxygen, and sulfur. On a practical and much larger scale, the utilization of isotope effects underlies the industrial production of heavy water and the separation of the fissionable isotope ^{235}U.

There are other anomalies that are attributable to radioactive decays. For example, ^{87}Sr has an abnormally high abundance in rocks rich in rubidium. This isotope is accumulated by the disintegration of radioactive ^{87}Rb. Significant variation in the isotopic composition of lead occurs in ores of uranium and thorium which decay into different stable isotopes of the element lead.

The confinement of a large number of positive protons (up to 92 in uranium) in such a tiny volume as that of the nucleus is incompatible with the familiar experience in the macroscopic world that charges of the same sign repel each other. The explanation for this must be sought at a deeper level in the structure of the nucleus.

NUCLEAR STRUCTURE

Mass and Energy

The external structure of the atom comprised of a cloud of electrons can be readily dismantled by an electric discharge or by

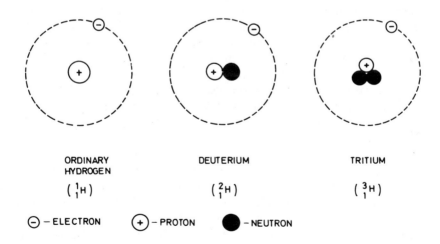

ORDINARY
HYDROGEN

$(^{1}_{1}H)$

DEUTERIUM

$(^{2}_{1}H)$

TRITIUM

$(^{3}_{1}H)$

⊖ – ELECTRON   ⊕ – PROTON   ● – NEUTRON

FIGURE 8. Isotopes of the element hydrogen.

the absorption of electromagnetic radiation which ejects electrons. Much more energy is required — at least a million times more — to expel a proton or a neutron from the nucleus. This energy corresponds to the nuclear binding energy. The mass of a nucleus is always smaller than the sum of the masses of all its nucleons and the missing mass is converted to nuclear binding energy.

The conversion of mass into energy is expressed mathematically in Einstein's relation $E = mc^2$, in which m is the mass and c^2 the square of the velocity of light. The transformation of mass into energy and, conversely, of energy into mass is routinely observed in a nuclear laboratory.

Because of the equivalence of mass and energy, both can be expressed in the same units. In nuclear physics it is more convenient to use energy units, and to express all masses in the equivalent quantity of energy. The energy unit is the electronvolt (symbol eV). It is defined as the kinetic energy which a singly charged particle (electron or proton) acquires in falling through a potential difference of 1 volt. On the nuclear scale it is a very small energy, and frequently used multiples are kiloelectronvolts (1000 electronvolts, 1 keV) and megaelectronvolts (1 million electronvolts, 1 MeV). With the mass-energy relation it is easy to calculate the energy equivalence of the constituents of the atom. The mass of an electron in energy units is 0.511 megaelectronvolts. This is the mass at rest: when the particle moves, its mass increases. For example, the mass of the proton at rest is equivalent to 938.26 megaelectronvolts: it increases 1000 times when the particle is accelerated with a kinetic energy of 10^{12} electronvolts.

The energy content of 1 kilogram of matter (rock, water, air) is enormous, and amounts to 5.6×10^{29} megaelectronvolts. Theoretically, less than 3 tons of matter could satisfy the annual world consumption of energy in the mid -1980s. But the conversion of matter into energy via common procedures like combustion or chemical or biological reaction has a vanishingly small yield. At the present time, the mass-energy relation has practical significance only in nuclear fission and in nuclear fusion.

The Forces

The structure and behavior of matter from the cosmic to the subnuclear level are inferred from four fundamental forces observed in the universe. These forces govern the interaction between all types of bodies from stars to atoms. Two such forces are well known in the macroscopic world; they are gravitational and electromagnetic. The other two, the strong force and the weak force, are typical for nuclear interaction.

An interaction can be viewed as an exchange of "quantas" (i.e., particles) between two objects, in the same way as two tennis players interact via the "quantum" tennis ball. For each type of force there is a quantum of the field force called gauge boson (Table 2). These force-carrier particles play the role of "mortar" in holding matter together.

Table 2

**THE FUNDAMENTAL FORCES IN NATURE AND
THE ASSOCIATED GAUGE BOSONS**

Force	Range of interaction	Relative strength	Gauge boson
Strong	Less than 10^{-15} m	1	Gluon
Electromagnetic	Very large	10^{-2}	Photon
Weak	Less than 10^{-18} m	10^{-13}	W^+, W^-, and Z
Gravitational	Very large	10^{-38}	Graviton

On the subatomic level, the behavior of atoms is governed by electromagnetic, strong, and weak forces.

The electromagnetic force holds the negatively charged electrons in their orbits around the positively charged nucleus and is the basis of stability in the atom. This is the Coulomb force, which has been recognized since the 19th century. Coulomb's law states that electric charges of like sign repel and charges of opposite sign attract each other with a force proportional to the product of their charges, and inversely proportional to the square of the distance between them. The force-carrier of electromagnetic interaction is the photon, a bundle of electromagnetic radiation whose energy can vary from very high (gamma-ray photons) to very low values (radio-wave photons).

The most powerful natural force is the strong force, which holds the atomic nucleus together. It is strong enough to overcome the enormous electromagnetic repulsion between the close-packed protons, and acts as efficiently between two protons as it does for two neutrons, or between a proton and a neutron. The strong force is over a hundred times more powerful than the electromagnetic force. The range is limited to the size of the nucleus, i.e., to less than 10^{-15} meter. The strong force is mediated by force-carrying particles called gluons.

Recognition of the weak force arose from studies of β-radioactivity. Two apparently strange things happen in the β-decay. First, the energy of the emitted electron is not constant but varies from zero to a maximum value. Further, the negatively charged electron escapes from the positive nucleus without being apparently affected by the presence of the electromagnetic and strong nuclear forces. The explanation is that, together with the electron, the nucleus emits an accompanying uncharged particle with a vanishingly small mass, the neutrino, which carries the seemingly missing energy of the radioactive decay.

The force carriers of the weak force are the W and Z particles. These are enormously heavy in comparison not only with the two other gauge bosons (Table 3), but with most other elementary particles as well (Tables 4 to 7). The weak force, together with the strong one, is responsible for nuclear reactions within the core of the Sun.

The most commonly operative force in the universe is the attraction between two objects, denoted by the gravitational force. The law of gravitation, formulated by Isaac Newton (1642—1727), states that two bodies attract each other with a force which is directly proportional to the product of their masses and inversely proportional to the square of the distance between them. The energy of gravitational interaction is extremely small, but the attraction operates over prodigious distances. Gravity governs planetary motion around the Sun, and holds together the innumerable bodies in the galaxies, the confederation of galaxies in clusters of galaxies, and the assembly of galaxy clusters in super clusters. Without gravity there would be no universe. Nevertheless, the gravitational force is so weak that it has no measurable action within individual atoms. The particle which ensures gravitational interaction, the graviton, has not yet been discovered.

The four universal forces appear to be too different in their effects to permit a simple description within a single concept. A great deal of effort is now being made by physicists to construct a theory for unifying the four forces. Electromagnetic and weak forces have already been successfully merged into an "electroweak" theory. So far, the "grand unified theory" that incorporates the strong force into the electroweak theory is not entirely satisfactory. The inclusion of gravity in a fully unified theory represents even more fundamental obstacles.

Table 3

PROPERTIES OF GAUGE BOSONS

Name or symbol	Mass	Lifetime	Charge	Spin
Photon	0	Stable	0	1
W (W⁺ or W⁻)	83 GeV	10^{-25}s	+1 or −1	1
Z	93 GeV	10^{-25}s	0	1
Gluon	0	Stable	0	1

The Elementary Constituents

Atomic and nuclear structures can be simply described using the three basic constituents: electron, proton, and neutron. The electron is a structureless object with a radius less than 10^{-18} meter and a mass of 9.1×10^{-31} kilogram. It carries electric charge, 1.6×10^{-19} coulomb. The proton has a charge equal in magnitude to that of the electron but of opposite sign, and its mass is 1.67×10^{-27} kilogram. It forms the atomic nucleus of hydrogen, the simplest chemical element, and is a constituent particle of all nuclei. Recent theories suggest that the proton may be radioactive, decaying with the extraordinarily long half-life of about 10^{32} years. The neutron is an elementary constituent of all atomic nuclei except that of normal (light) hydrogen. It is about 0.1 percent heavier than the proton and has no electric charge. Outside the nucleus, the neutron decays with a half-life of about 15 minutes into a proton, an electron (a β-ray) and a neutrino. More precisely, when a neutron (n) decays it converts into a proton (p) by emitting a W⁻ particle. This carrier of the weak force almost immediately converts into an electron (e⁻) and an antineutrino (v̄).

Besides charge, mass, and lifetime, elementary particles are characterized by a less tangible property known as spin. They behave like spinning tops which, however, can only spin at allowed rates which are specific for each kind of particle. The spin value is determined experimentally and is expressed in units of a universal constant called the Planck constant. The electron, proton, and neutron each have spin values of $1/2$ in these units, whereas the force-carrying particles (the gauge bosons) have spins of 1.

Experiments have revealed that the subatomic world consists of a large number of elementary constituents in addition to the electron, proton, and neutron. About a hundred particles and their varieties are known at the present time. Particle physicists believe that this number can ultimately be reduced to only three types of truly elementary particles: the gauge bosons, the leptons, and the quarks.

Among the force-carrier particles (Table 3) only the photons are a familiar concept to the layman. As electromagnetic radiation, they can be directly seen (as visible light), felt (as heat), or recorded (as γ-rays). The W⁺, W⁻ and Z particles can be observed indirectly as products of their radioactive decays in sophisticated experiments. The carrier of the strong nuclear force, the gluon, is a massless bundle of strong force in the same way that the photon is the massless bundle of electromagnetic radiation. Photons, however, can travel freely and indefinitely through space, whereas gluons are confined to the boundaries of a nucleon, i.e., 10^{-15} meter and their existence can be confirmed only indirectly. Evidence for their existence is based on the fact that a gluon produces a characteristic jet of particles which is distinct from that created by other elementary particles in collision processes.

No experiment has yet been conceived for the subdivision of leptons and quarks or for providing evidence of an internal structure. At present, they are considered as points of congealed energy which represent the true elementary constituents of matter in the universe.

"Lepton" is a collective name (from the Greek "small") for electrons, muons, neutrinos, and tau particles (Table 4). They react by means of electromagnetic and weak interactions, but are insensitive to the strong force. The best-known lepton is the electron. The muon and the tau are its heavier counterparts. The three types of neutrino have no charge and as yet their masses, if they exist, have not been detected.

Quarks are elementary particles which were first postulated in 1964 to explain the substructure of all elementary constituents other than leptons. At present six varieties — or "flavors" — of quark have been defined (Table 5). An unusual property is their fractional charge, which corresponds to one third or two thirds of the elementary charge carried by the electron and the proton.

It is considered that in the beginning of the evolution of the universe the quarks must have been very abundant, and one might ask whether all quarks ended up in nucleons or whether some "fossil" quarks may still remain concealed elsewhere. Estimates of their concentrations in natural samples indicate extremely small values, amounting to about 1000 per cubic centimeter of ocean water. So far, there is no evidence for the presence of free quarks in nature.

The strong force, which governs quarks, appears to bind them so tightly together that they cannot be dislodged with the presently

Table 4

THE LEPTON FAMILY

Name and symbol	Mass	Lifetime	Charge	Spin
Electron, e⁻	0.511 MeV	Stable	−1	$^1/_2$
Muon, μ	105.6 MeV	2×10^{-6}s	−1	$^1/_2$
Tau, τ	1.784 GeV	3×10^{-13}s	−1	$^1/_2$
Electron neutrino, v_e	O(?) or < 50 eV	Stable	0	$^1/_2$
Muon neutrino, v_μ	O(?) or < 0.5 MeV	Stable	0	$^1/_2$
Tau neutrino, v_τ	O(?) or < 70 MeV	Stable	0	$^1/_2$

Table 5

THE QUARK FAMILY

Name and symbol	Approximate mass	Lifetime[a]	Charge	Spin
Up, u	5 MeV	Stable	$+^2/_3$	$^1/_2$
Down, d	10 MeV	Variable	$-^1/_3$	$^1/_2$
Strange, s	100 MeV	Variable	$-^1/_3$	$^1/_2$
Charm, c	1.5 GeV	Variable	$+^2/_3$	$^1/_2$
Bottom (or beauty), b	4.7 GeV	Variable	$-^1/_3$	$^1/_2$
Top (or truth), t	>30 GeV	Variable	$+^2/_3$	$^1/_2$

[a] Quarks occur only in pairs (mesons) or as triplets (baryons); their lifetimes depend on the nature of the individual meson or baryon. The up quark, being the lightest one, is as stable as the proton which contains it.

attainable levels of laboratory-produced energy. Accordingly, the existence of quarks is based only on indirect observations.

The list of elementary particles also includes those corresponding to antimatter. Every elementary particle has its antimatter equivalent that is analogous in almost every respect except that its electric charge is reversed. The first antiparticle was already discovered in cosmic radiation in 1932. It was a positive electron or positron, and was later also observed in the radioactive decay of certain radionuclides.

When an antiparticle collides with its corresponding particle, they both disappear and their masses are transformed into energy. The annihilation of an electron and a positron releases energy which is twice the mass of electron, or $2 \times 0.511 = 1.022$ MeV. The energy of annihilation radiation in the event of a proton-antiproton collision is about 2000 times higher because of the correspondingly larger masses involved in the process. The products of annihilation reactions are photons, particles, and antiparticles. Today it is possible to create and examine antimatter particles in the laboratory in the same manner as for their material counterparts.

Pairs consisting of a quark and an antiquark form a group of elementary constituents called mesons (Table 6). For example, the positive pion (π^+) consists of a u quark and a $\bar{d}$ antiquark (written $u\bar{d}$).

Clusters of three quarks bound together by the strong force are called baryons (Table 7). All baryons have a mass equal to or greater than that of the proton. The most stable baryon is the proton itself. It consists of two u quarks and one d quark, which is represented as uud.

Table 6

PROPERTIES OF SELECTED MESONS

Name and symbol	Mass	Lifetime	Charge	Spin	Quark assembly*
Pion-plus, π^+	140 Mev	2.6×10^{-8}s	+1	0	u$\bar{\text{d}}$
Kaon-plus, K^+	494 MeV	1.2×10^{-8}s	+1	0	u$\bar{\text{s}}$
Upsilon, Y	9.46 GeV	10^{-20}s	0	1	b$\bar{\text{b}}$

* $\bar{\text{d}}$, $\bar{\text{s}}$ and $\bar{\text{b}}$ are antiquarks.

Table 7

PROPERTIES OF SELECTED BARYONS

Name and symbol	Mass	Lifetime	Charge	Spin	Quark assembly
Proton, p	938.3 MeV	Stable (?) or >10^{32} years	+1	$^1/_2$	uud
Neutron, n	939.6 MeV	stable in nuclei; 14.9 min in the free state	0	$^1/_2$	ddu
Lambda, λ	1.115 GeV	2.6×10^{-10}s	0	$^1/_2$	uds
Omega minus, Ω^-	–1.672 GeV	8×10^{-11}s	–1	$^3/_2$	sss

A general picture of the structure of matter is complex even when only the principal elementary constituents are taken into account. The descending levels of subatomic matter are schematically represented in Figure 9.

Experimental Approaches

"Particle physics" is a branch of nuclear science which deals with the properties of elementary particles. Production of the latter is based on the mass-energy relation $E = mc^2$, whereby energy can be converted into mass. The higher the available energy, the heavier the particles which can be produced.

The creation of the electron, the lightest particle, is easily achieved. Many radionuclides, or nuclear reactions performed with small accelerators, can provide photons with an energy above the threshold of 1.022 megaelectronvolts (twice the electron mass) required to produce the electron-positron pair. However, at least 2 gigaelectronvolts (2×10^9 eV) are necessary to produce a proton-antiproton couple.

A common way of revealing the structure of a nucleus and producing particles is to bombard a target nuclide with an

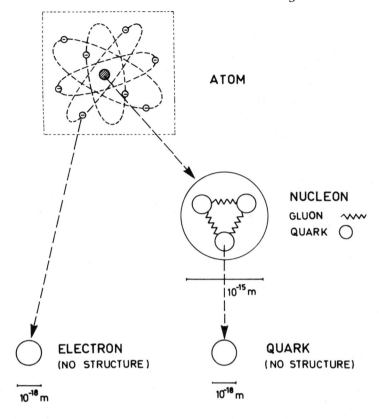

FIGURE 9. The descending levels of subatomic matter depicted schematically.

appropriate projectile. When the incoming energy is sufficiently high, the nucleus is broken up and tens to hundreds of particles are ejected from the collision site. They are detected in an indirect manner by their tracks in special devices.

The first "showers" of elementary particles were observed with natural projectiles, the cosmic protons. Coming from remote regions of the galaxies, protons with energies often much higher than those produced on earth collide with atmospheric nuclei, in the same way that protons strike a target in an accelerator. However, the observation of a cosmic shower is a matter of luck and the progress being made, although significant, is rather slow.

The artificial acceleration of protons started in the 1930s with very simple electrostatic machines producing potentials up to 1 million volts. The technology of both linear and circular accelerators has undergone continuous improvement. From a few hundred megaelectronvolts in the 1950s, the energy has been raised to the gigaelectronvolt range (equivalent to the proton mass) and now attains hundreds of gigaelectronvolts. The next generation of teraelectronvolt (= 1000 gigaelectronvolts) machines is planned or in construction. The Tevatron located at the Fermi Laboratory (U.S.) already accelerates protons at 1 teraelectronvolt; with that energy, the speed of the particles is very close to that of light (99.99995 percent).

When a high energy proton hits a target nucleus at rest, a large part of the energy appears as kinetic of the expelled particles. Thus the "internal" energy available for the production of matter and antimatter is rather low. The situation is quite different in a collision between two particles rapidly approaching each other, in which case the total kinetic energy can be converted into mass. These considerations have prompted the design of collision "rings" in which beams of a particle and its antiparticle are oriented in opposite directions.

At the European Center of Nuclear Research (CERN) in Geneva, protons and antiprotons are accelerated at 270 gigaelectronvolts. When the two beams intersect, a total energy of 540 gigaelectronvolts is available for the production of elementary particles.

The Super Proton Synchrotron at CERN is still one of the most powerful accelerators in Europe and provides protons of 400 gigaelectronvolts for fixed target experiments. In spite of its diameter of 2.6 kilometers it is a mere toy when compared to the 27 kilometer circumference of the superaccelerator, the Large Electron-Positron (LEP) collider, which can accelerate electron and positron beams up to 100 gigaelectronvolts. Thousands of the W and Z particles can be produced daily with this machine.

Direct experiments for testing the "Grand Unified Theory" are excluded. Even a circular accelerator with a diameter equal to that of the Earth would only provide a thousand millionth of the energy necessary to verify the model. However, the theory can be checked indirectly by its consequences, namely, that protons are not stable but exhibit an extremely weak radioactivity, with a mean lifetime of about 10^{32} years. Such a vanishingly small activity is equivalent to about 60 disintegrations per year in 1 million kilograms of matter. Therefore, the detectors of the decay products of the nucleons must be well shielded from cosmic radiation, which is much more intense.

For this purpose several laboratories have been installed at great depths in mines or tunnels. In Europe they are located in the tunnel of Mont Blanc in Italy and in the tunnel of Modane in France at a depth equivalent to 5 kilometers of water. The most common decay of a proton should be to a positron (e$^+$) and a pion-zero meson ($\pi°$), the latter decaying in turn to 2 gammas. It seems that by the end of the 1980s a few decay events have been observed, but they are too rare for enabling conclusions to be made.

The detection of neutrinos from the Sun and remote sources in the galaxy is also achieved in deep underground laboratories. The probability that these particles will interact with matter is extremely small. For 1 million neutrinos traversing the entire earth, only one has a chance of interacting and being detected. Again, the experimental setup must be well protected from cosmic radiations. Neutrinos are detected from the radioactivity which they induce in huge tanks containing compounds of the element chlorine. The method is based on measuring radioactive argon-37 which is produced after chlorine-37 captures the neutrino:

$$\text{}^{37}_{17}\text{Cl} + \nu \rightarrow \text{}^{37}_{18}\text{Ar} + e^-$$

Only a few atoms of Argon-37 (half-life 35 days) are formed at saturation in 400,000 liters of perchlorethylene (Chapter 5).

UNSTABLE NUCLEI AND RADIOACTIVITY

The stability of a proton-neutron configuration is the result of a subtle balance between opposite effects within the nucleus. To a first approximation, the binding energy of a nucleus is proportional to the number of nucleons. This suggests that each nucleon interacts with only a limited number of others, just as in a liquid where each molecule is bound to only a few neighbors. This analogy has led to the simple model which represents the nucleus as a liquid drop, in which the binding energy is considered to be due to both volume and surface effects. Nucleons at the surface have no neighbors on the outer side, and thus are more loosely bound than the inner nucleons. As the number of protons increases, the effect of the electric coulomb repulsion, a long-range force, becomes increasingly important with respect to the short-range attraction due to nuclear forces. Opposite trends are equilibrated in about 300 configurations, which are known as stable nuclei and are found in nature.

The total binding energy of a nucleus and, hence, its stability can be expressed as the difference between the sum of the masses of all constituent nucleons and the measured mass of the nucleus considered. The higher the difference, i.e., the loss of mass, the higher the binding energy. It is customary to express the average binding energy of the nucleons in a given nucleus. The nuclear stability is not uniform throughout the series of elements in the periodic system. It reaches a maximum of about 8 megaelectron-volts per nucleon for the elements around iron. Nuclei with a mass number smaller than 30 are less stable; similarly, the heaviest nuclei with atomic masses greater than 180, especially uranium, have a lower degree of stability (Figure 10).

With the exception of hydrogen, all nuclei have at least one neutron for each proton. Nuclei with an even number of protons and neutrons are largely predominant. It follows that nucleons have a propensity for grouping in pairs; this trend is strikingly reflected by the abundance and the exceptional stability of nuclei with mass numbers which are a multiple of four, i.e., that of the helium nucleus which consists of two neutrons and two protons. The numbers of nucleons, corresponding to an increased stability of the nuclear structure, are called "magic" numbers. Double magic nuclei, with magic numbers of protons and neutrons, are the most stable; such is the case for $^{16}_{8}O$ and $^{40}_{20}Ca$.

The most stable nuclides are also those which are the most abundant in nature. The element oxygen consists mainly of the

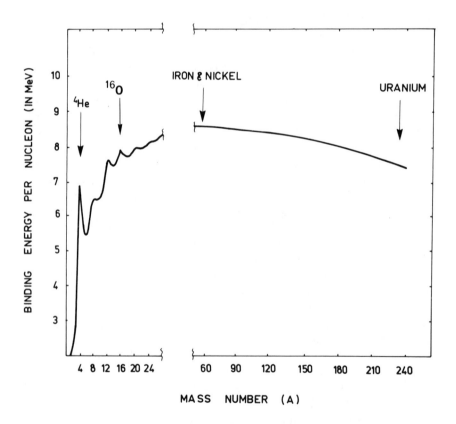

FIGURE 10. Average binding energy in stable nuclei.

isotope $^{16}_{8}O$, with a very small proportion of $^{17}_{8}O$ and $^{18}_{8}O$; the element calcium contains 96.9 percent of $^{40}_{20}Ca$.

Nuclear stability diminishes when the number of nucleons changes with respect to the stable configuration. Adding an excess of neutrons or protons to a stable nucleus, or removing one or two neutrons or protons is a way of making unstable nuclei. The spontaneous trend from instability to stability is a rule that governs the universe. Unstable nuclei undergo spontaneous transformation to recover a stable combination of neutrons and protons and this tendency constitutes the phenomenon of radio-activity. Unstable nuclei are radioactive.

Radioactive nuclei can be made artificially by nuclear reactions, which change the number of nucleons from that of a stable nucleus. For this purpose, nuclear reactors and accelerators are used. Nuclear reactions also occur in the universe and have led to the atoms which constitute our present world. Some of the nuclei formed in the early universe and of the solar system have not yet reached a stable configuration: they represent the primary natural radioelements such as uranium and thorium.

Modes of Radioactive Decay

The pathways from unstable to stable nuclei depend on the neutron-proton combination in the radioactive nucleus. A nucleus which has an excess of one neutron relative to the stable configuration decays in such a way that it will get rid of that neutron. This is accomplished by the transformation of the neutron into a proton. To preserve neutrality, an electron is emitted at the same time. This electron possesses all the properties of an atomic electron. The energy released in the transformation is carried by the electron as kinetic energy. For that reason, the electron which is expelled from the nucleus is called a β-ray or β-particle. In most cases, the energy of the β-particle is of the order of 2 to 3 megaelectronvolts, a much higher value than that of electrons in the atomic shell.

The escape of the negatively charged electron from the highly positively charged nucleus is explained by the theory of the weak interaction. Electron emission results from decay of neutron to give proton, beta-ray, and antineutrino:

$$n \rightarrow p + \beta^- + \bar{\nu}$$

In the β-decay, the daughter nuclide gains one proton, and thus its chemical identity is changed because the atomic number

increases by one unit. For example, radioactive carbon transmutes into nitrogen:

$$^{14}_{6}C \rightarrow \, ^{14}_{7}N + \beta^-$$

Similarly, radiophosphorus decays to give sulfur, radioiodine
gives xenon, and, contrary to the dreams of the alchemists,
radiogold transmutes to mercury. The lightest β-emitter is tritium,
the radioactive isotope of the element hydrogen. The daughter
nuclide of tritium is helium 3.

Radioelements which decay by emission of a β-particle are very
numerous. They are produced in large amounts in nuclear
reactors.

Some nuclides do not have enough neutrons to be stable. In
this case, a proton is transformed into a neutron. Charge conservation requires the simultaneous appearance of a particle carrying one unit of positive charge, the positive electron or positron.
Positron emission results from the decay of proton to give neutron,
positron, and neutrino:

$$p \rightarrow n + \beta^+ + \nu$$

The nuclide produced in the decay has lost a proton and has
changed its chemical identity. The atomic number decreases by
one unit. For example, radioactive sodium transmutes into neon:

$$^{22}_{11}Na \rightarrow \, ^{22}_{10}Ne + \beta^+$$

An alternative way of stabilizing a radioactive nucleus which
has a deficiency of neutrons involves the electron capture process,
whereby a proton in the nucleus combines with an atomic
electron:

$$p + e^- \rightarrow n + \nu$$

For instance, a radioactive isotope of iron is transformed into an
isotope of manganese:

$$^{55}_{26}Fe \rightarrow \, ^{55}_{25}Mn$$

Radioisotopes which decay by positron emission and by
electron capture are mainly produced in accelerators such as
cyclotrons, using charged particles as projectiles.

Heavy nuclei with mass numbers higher than 150 can disintegrate by emission of an alpha-particle. The latter is a nucleus of

helium ($^4_2$He) containing two neutrons and two protons tightly bound together. The kinetic energy of the α-particle is several megaelectronvolts.

A nuclide formed by α -decay contains two protons and two neutrons less than in the parent nuclide. Thus, the atomic number decreases by two units and the mass number by four units:

$$^{238}_{92}U \rightarrow {}^{234}_{90}Th + \alpha$$

Emission of α- and β-particles is a principal mode by which a radioactive nuclide attains a stable configuration. In recent years, new types of radioactivities have been discovered by emission of particles much heavier than helium, such as carbon or neon nuclei. Such events are very rare. A few nuclides are also known which disintegrate by emission of protons.

Heavy nuclei like uranium can also decay by splitting into two lighter nuclei: this phenomenon is known as spontaneous fission. It is only infrequently observed for uranium, but competes increasingly with α-emission when the atomic number increases. For the heaviest nuclei produced artificially, it represents the predominant mode of decay (Figure 11).

Gamma Rays

Nuclei, whether stable or radioactive, generally exist in the ground state. This is the state in which they have the smallest amount of energy. The same holds for atoms and molecules. All these species can also exist with a surplus of energy, in excited states. Excited species have a more or less pronounced tendency to return to the ground state and release the excess of energy by emission of photons. In the case of nuclei, these photons are γ-rays. They are analogous to X-rays emitted when atomic electrons jump from an outer to an inner level, but generally carry much more energy. In comparison with α– and β-rays, γ-rays have fairly long ranges in air and in materials of low density.

In many cases, radioactive decay does not produce the daughter nuclide in the ground state, but in one or several of the many accessible excited states. The lifetime of excited states varies within a very broad range, from 10^{-9} second or less to millions of years. Long-lived excited nuclei can be considered as radionuclides decaying by γ-emission. Parent and daughter have the same number of protons and neutrons; they differ only by their energies and are termed as nuclear isomers. This phenomenon is called nuclear isomerism. The couple of an excited and ground state of

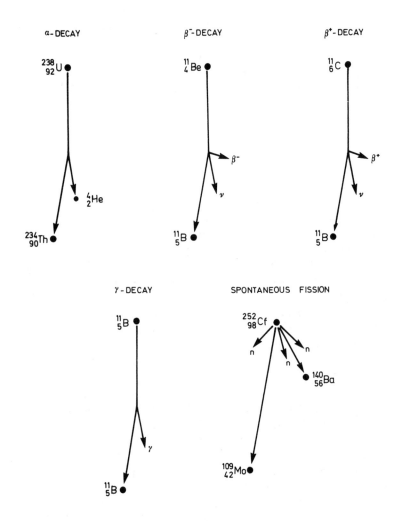

FIGURE 11.   Principal modes of radioactive decay.

a nucleus constitutes an isomeric pair, and the corresponding decay is an isomeric transition. The excited level is distinguished by the small letter m, for metastable. ^{80m}Br is the metastable isomer of ^{80}Br to which it decays with a half-life of 4 hours.

Probability of Radioactive Decay

Radioelements decay with very different rates. The half-life which characterizes the time for the radioactivity of a sample to fall to half its initial value is meaningful only for a large number of radioactive nuclei. The time at which a given radioactive nucleus will decay cannot be predicted. The decay of a nucleus is instantaneous and may occur at any time of observation during the first instant or after hundreds of years. No prediction can be made about the survival of a nucleus during the next second to come, in the same way that nobody can foretell the outcome of one trial when playing at heads or tails. Both are games of chance: head or tail, decay or nondecay. The probability that a given event will occur can be derived only by a mathematical treatment such as is currently used for random processes.

The probability that a nucleus will decay within a given time interval is the same for all nuclei of any radioelement. With a fair approximation, this probability is a constant parameter which is characteristic of the radioelement and independent of the physical and chemical state. This parameter is expressed in terms of the decay constant λ (lambda) which is the fraction of the radioactive nuclei present that will decay in a given unit of time. A high value of the decay constant means a fast decay rate and, consequently, a short half-life; a low value implies a high chance of survival, and thus a long half-life. The total number of decays per unit time is given by the product $\lambda \times N$ where N is the total number of radioactive nuclei at any given time. This product also defines the activity of a radioelement. In practice, the unit of time is the second.

Activity is a property quite different from other characteristics of a substance like length, weight, or temperature. Repeated measurements of the latter give a set of data grouped around the true value, with a degree of precision depending on the performance of the instrument and the skill of the experimentalist. This is not the case with activity measurements. Successive determinations will not bring the same result — or if so, only by chance. The experimental values fluctuate around the true value which can never be attained because of the statistical nature of the decay. However, for such a random process, one can estimate the chance that the true value will lie between two limits. To obtain a higher precision, the limits between which the "true value" will fall must be extended. For a single measurement with the result of, say, 10,000 becquerels, the true value has 68 percent chance of falling between 9,900 and 10,100 becquerels. The probability is 99.7 percent that the true value will lie between 9,700 and 10,300 becquerels, which is a high degree of certainty, but the limits are rather large. The precision of the measurement increases with the number of events recorded.

Radioactive Equilibrium

Quite often the nuclide formed in radioactive decay is itself radioactive and engenders a granddaughter of the parent which, in turn, may also be unstable. Hence, a parent nuclide can be the progenitor of several generations of radioactive daughters, which together constitute a radioactive family. The chain ends when a daughter decays to form a stable nuclide. The parent, as the ancestor of the family, has a longer half-life than all its descendants; otherwise the family would become extinct. Three radioactive families are found on Earth. The parents are isotopes of uranium and thorium, and in all families the end product is an isotope of lead.

After a sufficient length of time, the activity of the daughter becomes equal to that of the parent. This situation characterizes a radioactive equilibrium. Now the daughter decays with the half-life of the parent, because it is continuously regenerated from the latter.

Radioactive equilibria occur in nature. ^{238}U disintegrates with a half-life of 4470 million years into ^{234}Th, which, in turn, decays with a half-life of 24.1 days. In about half a year, the daughter is in equilibrium with ^{238}U and the activity of the pair decreases with the half-life of the parent. The decaying thorium atoms are continuously replaced by new atoms at the same rate. However, if by some means (including a natural process) ^{234}Th is separated from its parent, it decays with its own half-life and vanishes completely within a few months.

In a large family, the time needed to reach radioactive equilibrium is determined by the half-life of the longest-lived daughter; it can take up to thousands of years. Physical and chemical events which remove one or the other of the members of the family disturb the equilibrium. Short-lived radionuclides with half-lives down to a few seconds or less are still found on our 4500 million year-old Earth, because they are in equilibrium with parents which have survived throughout the enormous geological time span.

The equilibrium state between parent and daughters is of importance in the application of radionuclides, in particular in medicine. Frequently short-lived radionuclides with half-lives in the minute or hour range are the most convenient. They can be used at distances far from the production site provided that a long-lived parent is available. Parent and daughter are dispatched in the equilibrium state. When required, the daughter is separated by a simple chemical operation and is ready for use. Then the daughter grows again in conjunction with the parent as fast as it decays, and once the equilibrium is again established, a new batch can be separated. Quite appropriately the parent nuclide is termed a "radioactive cow" from which the daughter is periodically "milked" (Chapter 8).

RADIOACTIVE CLOCKS

Attempts to return into the past beyond the memory of the oldest man, beyond history and beyond the creation of man rely more and more on the achievements of science. The time span of events in which historians, archaeologists, anthropologists, geologists, and astrophysicists are interested differ widely, from a few centuries to billions of years. Numerous methods are used in dating, since obviously the same technique cannot be applied to a piece of pottery, to a rock, or to a star. But there is one phenomenon which finds universal application in dating: radioactivity. It is the basis of radioactive clocks, which provide the only reliable method of measuring very long intervals of time and which are equally suitable for dating more recent events.

The probability of the decay of a radionuclide is independent of the age of the nucleus and is unaffected by heat, pressure, magnetic and electric influences, and, in fact, all external forces. The basis for all radioactive dating methods is this constancy of decay rate. The disintegration process of natural radioelements preceded the formation of the Earth and has continued immutably until the present day at constant rates and with perfectly known kinetics. From measurement of the present activity of a radioelement, the amount which existed at any previous time can be calculated in a straightforward manner. The time elapsed since the incorporation of a long-lived radioelement into a closed system can be inferred from the residual activity or from the amount of elements formed in the decay. Radioactive clocks are widely used for the dating of archeological and geological and cosmic materials.

Radiocarbon Dating

^{14}C, a radioactive isotope of the element carbon, is constantly produced in the upper atmosphere from atoms of nitrogen struck by neutrons that originate in cosmic rays. Its half-life is 5730 years, which is a very suitable value for many archaeological measurements. Radiocarbon atoms react rapidly with atmospheric oxygen, producing radioactive carbon dioxide ($^{14}CO_2$), which is uniformly mixed with ordinary carbon dioxide in air. During photosynthesis, both forms are absorbed by plants and, subsequently, they pass into the food chain of animals. Radiocarbon is also distributed throughout the hydrosphere and, consequently, all the living animal and vegetable world, the so-called biosphere, should be weakly radioactive owing to its presence.

Both the cosmic ray flux and the average production rate of 2.5 atoms of ^{14}C per second and per square centimeter at the Earth's surface appear to have been constant over the past 100,000 years. This implies that the influx of ^{14}C into all living organisms has been constant over a very long time span in comparison with the half-life of the radioisotope. As a result, the decay of radiocarbon in the biosphere is exactly compensated by its continuous incorporation. As long as an organism is alive, a constant rate of exchange with the atmosphere occurs and the number of ^{14}C atoms per unit mass of living matter is constant; in every gram of natural carbon, about 15 nuclei of ^{14}C disintegrate each minute. When the organism dies, however, the uptake of radiocarbon ceases and the amount contained decreases steadily by radioactive decay. Since the decay law is known it is an easy task to calculate from the remaining activity the time that has elapsed since the equilibrium was disturbed. The longer the time elapsed, the lower the radiocarbon activity will be, and this provides the basis of radiocarbon dating. If the specimen has an activity corresponding to 7.5 decays of ^{14}C per minute and per gram, it will be concluded that one half of the initial radiocarbon has decayed, and the animal or plant must have died 5730 years ago. This dating method has been checked by comparison with specimens of wood of known ages from the tombs of the Pharaohs, or with certain old trees whose age is well established by other techniques (Figure 12).

To measure the activity of a sample, the radioelement must be introduced into a detector in the form of a gas or a compound which is soluble in an organic solvent. On both instances it is first necessary to convert the carbon into carbon dioxide, which can only be achieved by burning a relatively large sample of the specimen to be dated. This is obviously not a reasonable procedure in the case of irreplaceable materials like parchments or other precious relics. The sensitivity of the method is limited by the very low counting rate that will be recorded if only a minute portion of the specimen is available. A typical operation requires observation of 10,000 disintegrations for ages in the range 5000 to 10,000 years. This corresponds to 1 to 10 grams of carbon and counting times of 1.5 to 15 hours.

A new technique has recently been developed for determining a large fraction of the total number of radiocarbon atoms, instead of counting only the relatively few atoms which decay during the measurement. For each decay per minute there may be millions of millions (10^{12}) or more radioactive atoms in the sample. In radiocarbon dating based on accelerator mass spectrometry all

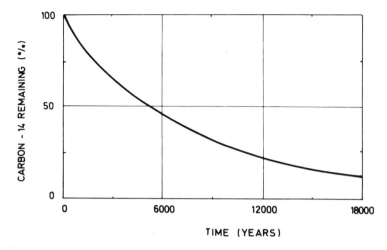

FIGURE 12. ¹⁴C as a radioactive clock.

¹⁴C atoms (not only the ones that disintegrate during the measurement) are separated and counted directly by a detector. Using the new technique, objects as old as 40,000 to 100,000 years can be dated with as little as 1 to 100 mg of carbon. Only 2 square centimeters of the cloth had to be sacrificed for the accelerator-mass spectrometer technique to show the medival origins of the Shroud of Turin (Italy), famous for its haunting image of a crucified man and regarded by some as the burial cloth of Christ. It gave 95 percent certainty that the linen of the Shroud dates from between A.D. 1260 and 1390, and virtually 100 percent certainty that it was made later than A.D. 1200.

The radiocarbon dating technique is applicable to any biological material whose death corresponds to a time of interest to archaeologists. Thousands of samples have now been dated with such significance that results from "preradiocarbon" datings have become obsolete. Charcoal, wood, bone, leather, textiles, ropes and fibers are among the materials which have been used to date archaeological sites.

Other Radioactivity Induced by Cosmic Rays

Other radionuclides with long half-lives are produced by cosmic rays and are potentially applicable to dating, in particular in combination with accelerator-mass spectrometry. Beryllium-10 is one of these. It is produced in the atmosphere at the rate of 0.015 atom per square centimeter per second near the Earth's surface, by the reaction of cosmic rays with oxygen and nitrogen nuclei. Its half-life is 1.5 million years, sufficient to allow uniform mixing of the radioelement with the atmosphere and the oceans. An equilibrium state is established in which the radioelement is deposited out on the ocean bed at the same rate at which it is produced. Incorporated in sediments, beryllium-10 enables the dating of very small rock samples for the younger specimens and requires up to a few hundred grams for the older material.

The shortest-lived radioisotope from cosmic ray sources that is used in dating is tritium, i.e. radioactive hydrogen. Despite its half-life of only 12.3 years, it has various applications in hydrology, meteorology, and oceanography.

Before the early 1960s, tritium in the atmosphere was produced only from cosmic rays, by the reaction of fast neutrons with oxygen and nitrogen atoms. Since then, most of the tritium arises from residues resulting from atmospheric testing of thermonuclear bombs. Artificially formed tritium behaves in the same manner as that of cosmic origin and this isotope is useful for dating bodies of water. Since the half-life of tritium is short, mixing over the surface of the Earth is incomplete. The tritium content of rainwater can vary widely, depending on geographical origin.

A typical example of tritium dating is the determination of time required for an underground natural water reservoir to be refilled after being tapped. Fresh samples of snow, ice, and even wines are accessible to tritium dating.

ISOTOPIC ANOMALIES AND EXTINCT RADIONUCLIDES

The abundances of the different chemical elements vary considerably within the solar system. On the other hand, with a few exceptions, the isotopic composition of elements in nature is extremely uniform. This proves that the elements of which the Earth, meteorites, and the Moon are formed were produced in the same nucleosynthetic process. By extrapolation, it may be assumed that the entire contents of the solar system, including the Sun and planets, were made from the same primordial material.

The few exceptions to the isotopic homogeneity result mainly from radioactive decays occurring in nature. Such is the case for lead formed from the decay of uranium and thorium. In minerals which were initially rich in these radioelements, the amounts of ^{206}Pb, ^{207}Pb, and ^{208}Pb formed by disintegration increased steadily, and the age of the mineral can be calculated from the isotopic composition of lead. Consequently, the lead isotope content of minerals will depend on that of uranium and thorium. For similar reasons, the isotopic composition of argon and strontium in natural samples depends on the amount of radioactive parent elements, potassium and rubidium, respectively, which were present at the time of formation of the rocks and minerals.

The elements in the periodic chart are classified in order of increasing atomic weights. Elements in a given column of the table have similar chemical properties. In order to preserve the chemical homogeneity of the classification, the positions of a few pairs of neighboring elements have to be inverted. For example, the atomic weight of argon is 39.948, while that of the adjacent element potassium is 39.098. This anomaly is due to ^{40}K, a long-lived natural radioactive isotope of potassium. The latter has a half-life of 1280 million years and decays to stable ^{40}Ar. The present abundance of ^{40}K is only 0.01 percent, but it was 12-fold higher when the Earth was formed. A large amount of ^{40}Ar from potassium was introduced into the atmosphere, where it mixed with the "normal" isotopes of argon which have mass numbers 36 and 38. Thus, the atomic weight of the rare gas has progressively increased up to its present value, which exceeds that of potassium.

A second inversion occurs for tellurium (atomic weight 127.60) and iodine (atomic weight 126.904). The latter comprises only one isotope, ^{127}I. However, at that time when the elements were synthesized in the universe, a heavier isotope, ^{129}I, existed in the same amount. The half-life of ^{129}I is 17 million years, which is relatively short in comparison with the 4500 million years of the Earth's existence. In accordance with the exponential decay of radioactivity, the amount of existent ^{129}I should now have decreased to about 10^{-80} of its initial value, which is equivalent to saying that the ^{129}I which existed when the Earth was formed has since completely disappeared. ^{129}I is an "extinct" radionuclide which has been completely transformed into stable ^{129}Xe. The extinction of the heavy iodine isotope is thus responsible for the lower atomic weight of this element with respect to tellurium.

All extinct nuclides have half-lives which are much shorter than 1000 million years. Many long-lived radioisotopes which are now made artificially were also formed during the early days of the universe, but by now have entirely disappeared. Such isotopes include ^{36}Cl (half-life 300,000 years), ^{26}Al (720,000 years), ^{53}Mn (3.7 million years), and many others.

Extinct nuclides are revealed by their stable daughter products, which can be used as chronometers for dating past events. For instance, ^{129}Xe has been found in abnormally high abundance in many meteorites. From this observation, it can be deduced that a significant quantity of ^{129}I must have existed in the meteorites, and hence also in the solar nebula when the meteorites were formed. From the existence of ^{129}I, it was concluded that the nucleosynthesis in the presolar nebula must have occurred only 100 million years before the birth of the solar system.

RADIOGENIC HEAT

The term "radiogenic" denotes any effect resulting from radioactive decay. When the radiations emitted by radionuclides are absorbed in matter, their energy is converted into heat. Radioactive materials are therefore warmer than their surroundings, and when the heat transfer is efficient the temperature of the environment rises. This occurs in celestial bodies and is achieved on an enormous scale inside the Earth, where the temperature has steadily increased since its formation as a result of energy liberated by radioactive decay.

Large amounts of radiogenic heat occur in nuclear reactors both during operation and shutdown periods. Nuclear fuel rods must be cooled when they are extracted from a reactor. However, the release of heat from currently handled radioactive sources is usually very small, e.g., 54.6 joules per hour for 1 curie of ^{60}Co. A dose of 10 kilograys (Chapter 3), provided by a multikilocurie source, would be required to increase the temperature of 1 gram of water by 2 degrees.

The rate of heat generation depends on the energy of the radiation and on the decay rate. The α-particles have a much shorter range than β-rays and γ-radiations and are also more energetic. Thus, the energy from α-particles is dissipated in a small volume and the heat density is considerably higher than in the case of other emitters, whose radiations are generally not fully absorbed in the radioactive material.

The heat released by radioactive elements can be converted into electricity: this is the principle of radionuclide batteries. The most important conversion mode uses thermoelements, which convert the heat into an electric current. Radionuclide electricity is costly and potentially dangerous because of the strong radioactive source, but it also has some advantages. It is very reliable, since nothing can stop the emission of radiations and the generation of heat. Its lifetime is determined by the half-life of the emitter. It must be neither too short nor too long, since an optimal power rate is desirable. The most important radioelements for radionuclide batteries are ^{238}Pu (α-emitter of half-life 87 years) used as plutonium oxide (PuO_2) and ^{90}Sr (β-emitter, half-life 28 years), in the form of strontium titanate.

These devices (referred to as SNAP, System for Nuclear Auxiliary Power) are particularly suitable when power is required over long periods in inaccessible location such as satellites, remote weather stations, and buoys.

For some time, an important use of radionuclide batteries was as a source of electricity in heart pacers. Such devices are required by patients whose natural cardiac rhythm is insufficient. The pacer emits pulses of electricity to maintain the heartbeat. For this purpose, the radionuclide battery must have a small volume (about 10 cubic centimeters) and contains up to 0.15 grams of ^{238}Pu. The battery is inserted into the patient's breast and need not be removed for at least 10 years. At present, however, pacemakers use newly developed, long-lived chemical batteries which avoid all risks connected with the implantation of high activities associated with hazardous α-emitters.

Further Reading

Radioactivity is treated in many elementary and advanced textbooks on nuclear physics and nuclear chemistry, for example:

Friedlander, G., Kennedy, J.W., Macias, E.S., and Miller, J.M., *Nuclear and Radiochemistry*, 3rd ed., John Wiley & Sons, New York, 1981.

Lieser, K.H., *Einführung in die Kernchemie*, 2nd ed., Verlag Chemie, Weinheim, 1980.

Vertes, A., and Kiss, I., *Nuclear Chemistry*, Elsevier, Amsterdam, 1987.

Aitken, J. , *Physics and Archeology*, 2nd ed., Clarendon Press, Oxford, 1974. [Includes radiocarbon and thermoluminescent dating.]

Roth, E., and Poty, B. (Eds.), *Nuclear Methods of Dating*, Solid Earth Sciences Library, Kluwer Academic Publishers, 1989. [An encyclopedic review of principles and applications of dating techniques.]

Close, F., Marten, M., and Sutton, C., *The Particle Explosion*, Oxford University Press, New York, 1987. [A remarkable up-to-date contribution to the popularization of modern physics and particle physics. Exceptionally well illustrated.]

In this chapter we consider basic physical, chemical and biological aspects of the interaction of ionizing radiation with matter and mention several applications. Nuclear reactions are discussed in Chapter 4.

Ionizing radiation loses energy when it passes through a medium and interacts with constituent atoms and molecules. The latter absorb energy and become excited or ionized, thereby generating chemically reactive species such as free radicals.

In representing the chemical action of radiation we consider the radiolytic behavior of water (liquid, vapor, ice) as a model system.

The biological consequences of irradiation are presented in the light of radiation chemistry of individual cell constituents as well as the findings at the cellular level. Exposure of the human body and the risks of irradiation are also considered.

Radiation detection and dosimetry are essentially based on physical changes induced by radiation and examples of related techniques are given.

Radiation-induced chemical and biological effects have found various applications in everyday life and ionizing radiation is beginning to emerge as a new, specific source of energy in industry.

3

Ionizing Radiation

NATURE AND ENERGIES

Emission of a ray, whether it be an electromagnetic wave or a particle, is called radiation. Interaction with the medium it traverses leads to degradation of the radiation energy and its dissemination in the medium. If energy absorption results in expulsion of electrons from atoms and molecules, the ray is called "ionizing radiation".

Ionizing radiations are energetic electromagnetic waves such as gamma (γ)-rays emitted by radioactive substances, or the penetrating X-rays which are produced by "roentgen" machines. Ionizing radiations also include the alpha (α)- and beta (β)-particles emitted in radioactive decay and various energetic, charged particles such as accelerated electrons, protons, and ions of heavier chemical elements produced in specially devised machines known as accelerators. Although neutrons in themselves are electrically uncharged, they may produce ionized particles in nuclear reactions or eject hydrogen ions from molecules which they strike.

In addition to ionization, energetic radiation can also cause excitation. This occurs when bound electrons acquire enough energy to reach higher energetic levels within an atom or molecule without being expelled. Excited atoms and molecules are very unstable and in condensed matter (such as a liquid or a solid) they usually release their excess energy within a small fraction of a second, as a quantum of electromagnetic radiation, the photon. The excited species can also react with neighboring atoms or molecules and such reactions are of paramount importance in photochemistry.

Atoms and molecules which are stripped of electrons during ionization become positive ions. Negative ions are formed when atoms or molecules acquire electrons in excess of those contained in the electrically neutral state.

The manner of formation and number of ionized species for a given medium will depend on the nature and energy of the radiation involved.

Electromagnetic Radiations

These are characterized by their wavelength or their energy: the shorter the wavelength, the higher the energy. Visible light is a mixture of various electromagnetic radiations, comprising wavelengths between 0.6 (red) and 0.4 micrometers (violet), for which the corresponding energies are 1.6 and 3.1 electronvolts. The energy required to ionize a light atom like hydrogen or oxygen exceeds 10 electronvolts, and thus visible light is not an ionizing radiation. It can only excite atoms and molecules.

Radiations with energies slightly higher than those of the visible range correspond to the ultraviolet rays, such as those emitted by mercury lamps or the Sun. Their energies are mainly in the range of 6 to 20 electronvolts.

Electromagnetic radiation with energies of the order of 1 kiloelectronvolt is denoted by the term "X-rays". These are produced when an electron jumps from a higher to a lower lying orbit in the atom, i.e., to one situated closer to the nucleus. A common practical method of producing X-rays is by bombardment of a metallic target with energetic electrons. The flux of X-rays emitted depends proportionally on the intensity of the electrons used as projectiles. Commercial X-ray tubes can produce thousands of millions of X-rays per second; their energy is determined by that of the bombarding electrons and on the type of metal used as target. For applications in medical radiography, the energies of electrons striking a tungsten anode are usually up to 250 kiloelectronvolts.

The absorption of X-rays depends on the density of a material and on the atomic number of the constituting elements. Thus, materials which have low values for these properties will be more transparent to X-rays. Accordingly, the muscles, or cracks in a bone, are more transparent than the bone itself and this permits a differentiation of contrast in a "radiograph".

Nowadays the old "roentgen" machines are being progressively replaced by sophisticated devices for tomography which enable photography of specific layers inside the body. In combination with computer treatment of data, this technique provides extraordinary possibilities for three-dimensional imaging of internal parts of the body.

For certain industrial applications such as controls of welded joints, or in medical cancer therapy, X-rays in the multikiloelectronvolt range are required. For such purposes the betatron is used. In this device, electrons are accelerated to megaelectronvolt energies and the energetic X-rays are generated when the electrons strike a metallic target.

X-rays of low energy but with enormous fluxes (about 10^{14} rays per second) are available in the so-called synchrotron radiation. The synchrotron is a device used for accelerating electrons up to energies of thousands of megaelectronvolts. Under the effect of a strong magnetic field these electrons emit an intense radiation in the far ultraviolet and X-ray regions. From the abundant fluxes produced, it is possible to select a very intense beam of X-rays with a definite wavelength. Synchrotron radiation provides a powerful tool for research in solid-state physics and X-ray crystallography. Whereas X-rays from conventional roentgen tubes enable the determination of atomic positions in a crystal lattice, synchrotron radiation combined with a technique known as extended X-ray absorption spectroscopy makes possible the investigation of the local environment of an atom in a complex molecule such as a protein. Although expensive, sources of synchrotron radiation are in routine use in several laboratories throughout the world. The construction of a giant synchrotron with a diameter of 500 meters for providing 60-gigaelectronvolt electrons for synchrotron radiation is being jointly prepared by several European countries and is expected to be put into operation in the 1990s.

γ-Rays are also electromagnetic radiation similar to X-rays but their origin is different: they are emitted during nuclear processes and not in electronic transitions. Radioactive elements are the most common sources of γ-radiation, for which energies lie in the kiloelectronvolt to megaelectronvolt range. For industrial application and medical therapy, radioisotopes emitting high-energy γ-rays and having long half-lives are required. The most common radioisotope for these purposes is ^{60}Co, whose half-life is 5.27 years and which has γ-ray energies of 1.33 and 1.17 megaelectronvolts. A kilocurie source, which is suitable for medical therapy but usually insufficient for industrial uses, emits 7.4×10^{13} γ-rays per second.

Charged Particles

Radioelements which decay by β-emission provide useful sources of negatively charged particles, i.e., electrons. Because of their short ranges, these electrons are used only for surface irradiations. A common β-emitter is ^{90}Sr (half-life 28 years, 0.55 megaelectronvolt β-particles). Its daughter nuclide is ^{90}Y (half-life 64 hours, 2.28-megaelectronvolt β-particles). This pair is a convenient source of 2.28-megaelectronvolt electrons, since the radioactive equilibrium between ^{90}Y and ^{90}Sr is established already within a few days following manufacture of the strontium source.

The kinetic energy of a particle depends on its velocity, and this can be increased by devices designed to accelerate charged particles (electrons, protons or nuclei of various atoms) with the aid of an electric field. Various types of such accelerators are currently in use.

The first electrostatic accelerators were based on the application of a high voltage, as in the case of the so-called Van de Graaff generators that are still being produced and used (Figure 13). The particles to be accelerated are produced in an ion source by stripping off one or more electrons from the atoms by means of an electric discharge in a gas such as hydrogen or helium. The ions are injected in a tube in which the air has been evacuated to a very low pressure, down to less than a millionth of the atmospheric pressure. This is required to avoid the loss of ions by collision with molecules of oxygen and nitrogen. The high voltage between the extremities of the tube provides the acceleration. Positive ions rush towards the negative electrode. The current of accelerated ions is usually in the microampere-milliampere range. Recently, special ion sources have been designed to produce beams of high intensity, i.e., up to several amperes, and these are used in the production of new materials by a process called "ion implantation".

In multiple-stage accelerators a lower voltage is applied repeatedly, in successive increments, until a high particle energy is finally attained. This is achieved in linear accelerators by a multiplication of basic components which results in a device extending in length. For production of protons with energies of several hundred megaelectronvolts, or electrons with 10,000 megaelectronvolts or higher, the lengths of linear accelerators may reach several hundred meters.

In studies of fast physical-chemical processes induced by radiation, a special type of electron accelerator is used which

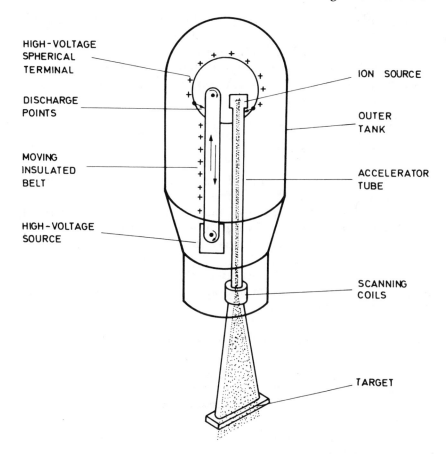

HIGH-VOLTAGE
SPHERICAL
TERMINAL

DISCHARGE
POINTS

MOVING
INSULATED
BELT

HIGH-VOLTAGE
SOURCE

ION SOURCE

OUTER
TANK

ACCELERATOR
TUBE

SCANNING
COILS

TARGET

FIGURE 13. Simplified diagram of a Van de Graaff accelerator; charges are transferred to the upper part by a conveyor belt.

delivers intense bursts of electrons. The pulses are of very short duration, often down to a picosecond (10^{-12} second). The pulse current is extremely high (about 1000 amperes), and induces a sufficiently large number of phenomena that studies can be carried out following the pulse and in the absence of radiation.

Neutrons

The extent of interaction of neutrons with electrons is so small that direct ionization is negligible. Indirect ionization, however, may be significant. The fate of neutrons is essentially determined in nuclear reactions, in which charged particles such as protons and alphas are often produced. These species, in turn, ionize atoms and molecules in the surrounding medium. In another type of indirect ionization, which is particularly efficient in a hydrogen-rich material like water, an energetic neutron ejects a hydrogen ion by collision, in the same way that an electromagnetic ray causes emission of an electron upon hitting an atom.

Nuclear reactors and other neutron sources are considered in Chapter 4.

PHYSICAL ACTION

The interaction of radiation with matter involves a transfer process in which energy is absorbed by the matter at the expense of the incident radiation energy. These processes are accompanied by subsequent changes in the radiation, as well as in the physical, chemical, or biological properties of matter.

Interaction begins with the physical action of radiation which, for a given medium of particular chemical composition and physical state, depends on radiation type and energy. It ends as radiation energy uptake.

Degradation and Deposit of Energy

Electromagnetic radiations at submegaelectronvolt energies transfer their energy to the medium mainly in a single event, which is the photoelectric effect. The photon imparts its energy to an electron, ejects it from an atom or molecule, and disappears (Figure 14A). In this event a single ion is formed, but the ejected electron in turn ionizes and excites many atoms and molecules. The energy of the incident photon is essentially transferred through the subsequent reactions of the photoelectric electrons.

In place of giving up its entire energy to a bound electron in a single event, a photon may transfer only a part and, after being deviated from its original path, continues to interact and lose further amounts of its energy. This scattering is designated according to its discoverer A.H. Compton, and the Compton electrons are mainly responsible for multiple ionizations. The Compton effect occurs at all energies, but is of particular significance at energies of about 1 megaelectronvolt (Figure 14B).

At still higher photon energies (several megaelectronvolts) another process becomes important, namely, pair production. This involves disappearance of the photon in a single event which gives birth to a positron and an electron. Pair production is always followed by a rapid annihilation of the positron and a subsequent formation of two γ-rays, whose energies (0.51 megaelectronvolt) are degraded mainly through the Compton effect (Figure 14C).

Ionizing Radiation **75**

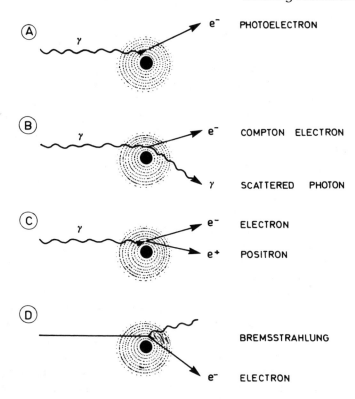

FIGURE 14. Degradation and deposit of energy of radiation. Photoelectric effect (A), Compton effect (B), pair production (C), Bremsstrahlung production (D).

At energies up to 1000 megaelectronvolts, positively charged particles such as protons or alphas also lose their energy mainly by interactions with electrons. In contrast with the single event occurring with electromagnetic radiation, energy loss in this case is continuous and results principally from energy transfer to electrons removed from atoms or molecules during collisions. Many of these secondary electrons which originate along the particle's pathway are sufficiently energetic to ionize further atoms and molecules. Up to 80 percent of the ionization produced by positively charged particles is due to this secondary radiation, which is termed "delta (δ)-rays".

The degradation mechanism becomes more complex at energies of about 1000 megaelectronvolts (1 gigaelectronvolt) or higher, as in the case of cosmic rays. In addition to loss due to reactions with electrons, very energetic charged particles also lose their energy by nuclear reactions and ejection of nucleons. The ejected nucleons subsequently interact with matter and undergo degradation of their own energy in cascade processes which produce further nuclear collisions and nucleon ejection. In nature, this gives rise to the cosmic rays showers. Their existence is a precious source of information on high-energy particle physics and events far beyond the Earth, but it also represents a potential danger for man during protracted periods of time in cosmic space.

Processes which are responsible for the energy loss of electrons are in many ways similar to those described for positively charged particles. The slowing down of an incident electron is almost entirely due to its interactions with electrons of atoms and molecules in the irradiated medium. The ejected electrons in turn ionize and excite the atoms and molecules along their pathways. In consequence, the role of primary ionization is subordinate in this case also and secondary ionizations contribute up to 80 percent of the total effect.

When electrons with higher energies traverse a relatively dense medium, the slowing down is accompanied by an emission of X-ray called "bremsstrahlung" (German, meaning "brake radiation"). This results from direct conversion of the kinetic energy of an electron into electromagnetic radiation (Figure 14D). Bremsstrahlung is also observed when β-rays from a radioactive source impinge on a shielding material. For heavier materials such as lead, the bremsstrahlung already becomes significant at electron energies above 1 megaelectronvolt. Accordingly, low-Z materials such as plastics are used for shielding β-emitting sources.

Penetrating Power

In a given medium, the range of radiation depends on its type and energy. Knowledge of this parameter is of great importance for applicational purposes and for radiation protection. Theoretical and experimental values are available for the ranges of various types and energies of radiations in a large number of materials.

Heavy charged particles travel relatively slowly, react very efficiently with neighboring atoms or molecules along the pathway and consequently lose their energy completely within very short distances. α-Particles from radioactive nuclides travel about 3 centimeters in air and only a few micrometers in body tissue. Very energetic cosmic protons may penetrate a layer of ice to a depth of 20 meters.

β-Rays travel faster, undergo fewer interactions, and give up less energy per unit length of pathway. Their ranges are correspondingly greater than those of alphas and amount to 3 meters in air and about 1 centimeter in body tissue. The range of a 5-megaelectronvolt electron in tissue is 2 centimeters.

γ-Rays travel at the speed of light, release relatively little energy per unit length of path and can cover large distances even in a dense medium. They pass through the human body, and a thick slab of concrete or lead is needed to absorb them; for γ-rays of ^{60}Co, a lead thickness of 15 centimeters is required to reduce the intensity by a factor of 5000.

Detection

Most detection devices are based on physical effects such as the ionization or excitation of a gas or solid.

Ionization in a gas produces an electron and a positively charged ion. If a potential of several hundred volts is applied between two electrodes, the charged species are displaced towards the electrode of opposite charge. The resulting flow of ions constitutes an electric current which is a measure of the intensity of the radiation in the gas, usually air. Such a system is known as an ionization chamber (Figure 15A). Its technical realization depends on the particular application. Thin walls are required for radiation of low penetrating power; the gas volume may be a few milliliters for personal instruments or several liters in specific cases.

The Geiger-Müller (GM) counter is a cylindrical ionization chamber filled with a mixture of gases (Figure 15B). The cylinder wall constitutes the negative electrode and a thin central wire carries the positive charge. If the voltage between these electrodes is high — about 1000 volts or more — a single ionizing particle will cause an avalanche of ionizations, thus leading to a very large pulse of current which can be easily measured. This permits detection of even low-intensity radiation. In practice, a GM counter consists of a glass or metal tube filled with a mixture of gases such as argon and methane; the choice of the wall material and gas depends on the type of radiation to be measured. The GM tube is often mounted in a convenient pocket-sized container which includes the power supply and electronic parts.

Solid-state semiconductor detectors are crystalline substances in which a phenomenon similar to ion-pair formation in a gas takes place. The photon or other particle releases an electron within the rigid structure and the resulting vacancy (called the "hole") is analogous to a positive ion in a gas. When an electric potential is applied, current flows as in the case of an ionization chamber filled with gas. However, the solid-state system has the advantage that it can be made much smaller and more sensitive than an ion chamber. Conductivity detectors frequently utilize cadmium sulfide.

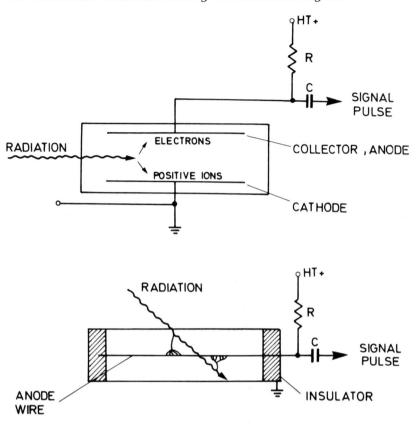

FIGURE 15. Basic principles of two common detection devices: the ionization chamber (A) and the Geiger-Müller counter (B).

In other solid-state detectors, radiation measurement is based on excitation rather than ionization. An electron is raised to a higher energy level but it still remains bound to the hole by electrical attraction. Its return to the ground state is followed by the release of excess energy as a pulse of light. The "scintillations" are detected by a photomultiplier tube, a device which converts the light pulse into electrical ones which provide information on radiation intensity and energy. A common scintillator for γ-rays is a crystal of sodium iodide (NaI).

Energies of ionizing radiation are measured with spectrometers, of which the semiconductor spectrometer is routinely used for analyzing γ-ray spectra. In this procedure, the radiation is fully absorbed in a suitable material and the energy converted into an electric pulse with a corresponding amplitude. The spectrometer also includes an electronic part and a computer system which sorts out the electric pulses from the detector and treats the information. The spectrometer provides data on energies and intensities of the radiation measured and enables identification of radionuclides and the evaluation of activity.

Fast neutrons are detected by their interaction with a hydrogen-rich material such as polyethylene, which can be incorporated into the ionization chamber. The neutrons eject protons, which in turn ionize the air in the chamber. For detection of thermal neutrons the chamber is lined with a thin layer of boron, an element which efficiently captures neutrons and releases α-particles.

The blackening of a photographic plate exposed to radiation is a detection method as old as the discovery of radioactivity, and is still widely used in nuclear physics, radiation dosimetry, and radiation imaging. Grains of silver bromide form the active part of the emulsion. Incident radiation acts on the silver bromide by producing electrons, which are trapped in the crystal grains and leave positive silver ions. The latter form a latent image that is developed into a visible image by chemical processing similar to that used routinely in photography. The tracks of individual particles left in the emulsion can be observed with a microscope; their lengths and densities provide information on the nature and energy of particles.

The film badge is the most common personal radiation dosimeter, and is worn by persons exposed to this risk. It consists of a small strip of film in a package which excludes light but allows passage of ionizing radiation. The amount of radiation is estimated after one or several weeks; the film is developed and the degree of blackening is compared with that of standards.

Autoradiography is a technique used for reliable and inexpensive imaging, for example in botany for visualizing the accumulation of labeled chemical species in certain parts of a plant. The specimen is kept in close contact with the film until a suitable degree of blackening is achieved.

Dosimetry

Medical doctors were the first to apply ionizing radiation in a practical sense; they termed the amount of energy absorbed the "radiation dose" and its measurement, "dosimetry".

The absorbed dose is the energy imparted by ionizing radiation to a unit mass of irradiated matter. Usually it is referred simply as the dose. The dose rate is the dose delivered per unit time.

The first internationally accepted recommendations on radiation units appeared in 1928. Since then two branches of radiation science have undergone continuous development, and these are dosimetry and health physics. Methods have been provided for the calculation of radiation doses and for measurement techniques, as well as for effective procedures in human protection and safe handling of radioactive materials.

Dosimetry standards are periodically revised, modified, and supplemented as new progress is made. Examples of present-day standards are given below.

In the International System of Units (Système International, SI), the energy is measured in joules and the mass in kilograms. Thus, the delivery of 1 joule in 1 kilogram of any material is the unit of absorbed dose of radiation and is called gray (symbol Gy). A former unit which is still frequently used is the rad (1 gray is equivalent to 100 rads).

Limitations are imposed by the use of gray in dosimetry of living matter. It represents the absorbed dose in the human body or other living system only when a particular type of ionizing radiation is considered, e.g., γ-rays, and when degrees of biological action of different absorbed doses are compared. When assessing the effect of mixed radiations of different types and energies, a total absorbed dose expressed in grays has little significance because of the different biological effects of the radiations involved.

In order to take this difference into account, the dose equivalent, expressed in sievert (Sv), is used in the SI. This unit also has the dimensions of joules per kilogram. For most penetrating radiations, including X-rays, γ-rays, and β-rays, 1 sievert is equal to 1 gray. However, for neutrons and α-particles, which have a higher relative biological effectiveness (RBE), values in grays are multiplied by a factor which depends on the type and energy of radiation; 1 gray is conventionally counted as equivalent to 10 sieverts if delivered by neutrons and 20 sieverts if referred to α-particles. A former unit which is still often used is the rem: 100 rems is equivalent to 1 sievert (Table 8).

The activity of a substance is expressed as the number of disintegrations per second (in becquerels); the dose is calculated by taking into account the energy released in each decay and the number of becquerels.

To promote reliable dose measurement various international programs have been set up. In the mid-1980's, one of them has assisted 650 hospitals worldwide through one dose-intercomparison service in the field of radiotherapy.

Imaging and Gauging

Roentgen already used the penetration of ionizing radiation to "see through" opaque objects such as parts of the human body.

Table 8

**UNITS OF MEASUREMENTS IN THE SYSTEME INTERNATIONAL
(SI UNITS) FOR IONIZING RADIATIONS**

Quantity	Name of unit	Symbol	Definition	Former unit	Conversion factor
Absorbed dose	Gray	Gy	Joule/kilogram	Rad	1 Gy = 100 rads
Dose equivalent	Sievert	Sv	Joule/kilogram	Rem	1 Sv = 100 rems
Activity	Becquerel	Bq	Second^{-1}	Curie (Ci)	1 Bq = 2.7 × 10^{-11} Ci 1 Ci = 3.7 × 10^{10} Bq

Other fields also exist in which the dependence of radiation penetration on the density of an absorbing material provides valuable information which would otherwise be inaccessible. For example, imaging techniques have been developed for finding cracks in metallic objects, locating unfilled cavities in metal castings or revealing unequal mixing of constituent metals in steel or alloys (Figure 16).

As opposed to stationary or mobile X-rays or ultrasonic units, which are often used for the same purposes, radioactive sources have the advantage that they do not require a power supply and are more compact and easier to operate under field or factory conditions.

Similar criteria hold for gauging. The amount of radiation which passes through a material provides a basis for a thickness gauge, whereby the thickness is related to the amount of radiation

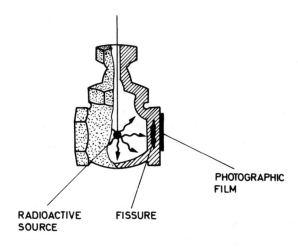

PHOTOGRAPHIC FILM

RADIOACTIVE SOURCE

FISSURE

FIGURE 16. Industrial radiography, a widely used nondestructive testing method.

which reaches the detector. The information thus obtained can be used for automatic regulation of a process. A large choice of isotopic sources with different radiation energies is available for the process control of a wide range of thicknesses. Many other such gauges are used for measuring materials in gaseous, liquid, or solid form, often under adverse conditions of temperature or humidity or in corrosive media (Figure 17).

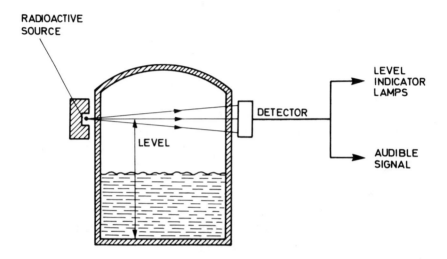

FIGURE 17. Level gauging. When the liquid in the tank reaches the level of the source, the signal received by the detector drops and triggers the alarm system.

At the present time, hundreds of thousands of various devices with radioactive sources are in use in industrial plants throughout the world. For example, the gauges continuously monitor the paper weight per unit area in the paper industry, or the thickness of steel to optimize the product quality in the steel industry. However, there is a recent trend toward replacement of this equipment by nonradioactive apparatuses because of the potential dangers associated with radioactive sources when they are used or stored in the absence of adequate controls.

CHEMICAL ACTION

Absorption of the energy of ionizing radiation by a substance may result in chemical changes. Decomposition due to radiation is called radiolysis; during this process free radicals are formed and these can lead to the formation of new products.

We shall take the radiolysis of water as a model system in presenting the basic aspects of the chemical action of radiation. Water is one of the simplest molecules and its radiolytic behavior is well understood for each of its physical states, i.e., in the form of vapor, liquid, or ice. This example is particularly appropriate also in view of the omnipresence of water and radiation in nature and the important role of water in biology and in various applications of nuclear energy (e.g., as solvent, reactor coolant, and protection material).

Radiolysis of Liquid Water

When a water molecule absorbs radiation energy, it becomes ionized:

$$H_2O \sim\sim> H_2O^+ + e^-$$

or excited

$$H_2O \sim\sim> H_2O^*$$

The symbol $\sim\sim>$ denotes a radiation chemical reaction in contrast with the symbol $\rightarrow$ used in chemistry. The asterisk * is used to designate an excited molecule.

The excitation energy is often sufficient to break the bond between one hydrogen atom and the rest of the water molecule, as represented by

$$H_2O^* \rightarrow H + OH$$

The fragments corresponding to the hydrogen atom (H) and the hydroxyl radical (OH) are chemically reactive species. If energy imparted in the ionization process is sufficient to cause an electron to escape the positive electric field of its parent ion, the H_2O^+ thus formed reacts with a water molecule to produce a hydronium ion and a hydroxyl radical:

$$H_2O^+ + H_2O \rightarrow H_3O^+ + OH$$

The escaping electron travels through the liquid, collides with water molecules, and loses energy by producing excitation and ionization of further water molecules. When it has been slowed down sufficiently to become trapped by neighboring water molecules, i.e., hydrated, a unique chemical species is formed, viz., the hydrated electron:

$$e^- + nH_2O \rightarrow e^-_{aq}$$

This is a quasi-entity in which one electron is shared by an aggregate of several water molecules, probably six. It is the simplest of negative ions and behaves like a normal monovalent ion.

The hydrated electron has a very strong tendency to become attached to atoms and molecules, which are then reduced, since e^-_{aq} is the strongest chemical reducing agent known. Because of its high reactivity its lifetime is very short, being of the order of microseconds in a condensed medium. Its discovery in the early 1960s was an important contribution of radiation research to chemistry.

The hydrogen atom is another reducing agent which is produced by the radiolysis of water. It is produced in lower yield and is less reactive than the hydrated electron.

The hydroxyl radical is a strong electron acceptor and its chemical behavior is characteristic of that of an oxidizing species.

The hydrogen atom, hydroxyl radical, hydronium ion, and hydrated electron are considered as the primary species of water radiolysis. They deviate from their radiation tracks while reacting with each other; among the dozens of reactions which are well established, of particular importance are the reformation of water,

$$H + OH \rightarrow H_2O$$

and the reactions between identical species which produce molecular hydrogen (H_2) and hydrogen peroxide (H_2O_2):

$$H + H \rightarrow H_2$$
$$OH + OH \rightarrow H_2O_2$$

A very efficient reaction also takes place between the hydrated electron and a hydronium ion, whereby atomic hydrogen is formed:

$$e^+_{aq} + H_3O^+ \rightarrow H + H_2O$$

These reactions are complete within 1 nanosecond (10^{-9} second) after the passage of ionizing radiation, and the overall chemical action of radiation on water can be summarized as:

$$H_2O \dashrightarrow H, OH, e^+_{aq}, H_2, H_2O_2, H_3O^+$$

Radiolytic products are located in the vicinity of the radiation pathways and diffuse away by reacting among themselves or with substances present in water. Irradiation of water of standard purity, for example, ends up as a very dilute aqueous solution of molecular hydrogen and hydrogen peroxide with traces of compounds produced by reactions of minute amounts of impurities with free radicals (H, OH, e^-_{aq}).

Radiation Chemical Yields

The extent of chemical changes depends on various factors. A convenient expression is the radiation chemical yield i.e., the number of chemical species (molecules, ions or free radicals) which are decomposed or formed for each 100 electronvolts absorbed. For example the radiation chemical yield of the decomposition of water by γ-rays is 4.5 which means that 4.5 water molecules are decomposed for every 100 electronvolts absorbed in liquid water.

A common radiation chemical yield is about 5. Occasionally it may be higher, particularly in the case of chain processes such as polymerization in which the yield number may exceed 1000. A few rare compounds are stable to radiation or their change is negligible even at large doses.

The same primary species are formed in water regardless of the type and energy of radiation but radiation chemical yields may vary significantly. For example, considerably more molecular hydrogen and hydrogen peroxide are produced in water by 4.8-megaelectronvolt alpha particles from ^{226}Ra than by 1.7-megaelectronvolt β-radiation from ^{32}P. The ranges in water of these particles are very different, being 33 micrometers for α-particles and 8000 micrometers for β-particles, which explains the difference observed in radiation chemical yields: the α-particles transfer much more energy within a shorter pathway. As a result, the primary events (excitation, ionization) and the primary species (hydrogen atom and hydroxyl radical) are more highly concentrated within the tracks of α-particles. Thus, their combinations are enhanced and the formation of molecular hydrogen and hydrogen peroxide is increased.

The higher yield of molecular hydrogen and hydrogen peroxide leaves less hydrogen atom and hydroxyl radical for chemical reactions. The enhanced reformation of water (H + OH) results also in a smaller number of free radicals available for oxidations (OH) and reductions (H, e^-_{aq}). For example, the chemical yield of oxidized iron is only 5 for 4.8-megaelectronvolt α rays, but up to 15 for the β rays of ^{32}P which have a much lower ionization density.

Spatial Distribution of Events

The distribution of primary events along a radiation pathway can be visualized by comparing it with a string of beads. Ionizatons and excitations are said to lie within a "bead" or a "spur". Estimates of distance between adjacent "beads" can be derived from known data on radiation energies and their ranges in water. In the case of 4.8-megaelectronvolt α-particles, the mean distance between beads is only 0.001 micrometer (10^{-9} meter), whereas it is about 0.5 micrometer (5×10^{-7} meter) for 1.7-megaelectronvolt β-rays of ^{32}P. When the distance is 0.001 micrometer as in the case of α particles, the beads overlap and the track can be considered as a cylinder (Figure 18).

The spatial distribution of reactive species influences the early events of radiolysis and the subsequent reaction mechanism. The "packing density" of species and their reaction products depends on the rate of energy loss, which is generally expressed in terms of the "linear energy transfer" (LET). This represents the local deposit of energy resulting from passage of an ionizing radiation through a medium.

An average value of the LET is calculated by dividing the energy of the particle by its mean range and is often expressed in kiloelectronvolts per micrometer. The LET for water of α-particles from radium is 145 kiloelectronvolts per micrometer, and that of β-particles from ^{32}P is 0.21 kiloelectronvolt per micrometer.

Thus, for a given medium the LET values depend on the type and energy of radiation. Numerically, for water they range from about 0.2 kiloelectronvolt per micrometer for penetrating radiation such as fast electrons or γ-rays of ^{60}Co, to several hundred kiloelectronvolts per micrometer for the fission fragments of uranium.

The average LET values are only approximative, since some of the calculational parameters are neglected for the sake of simplicity. In reality, the true rate of energy loss is not constant throughout the range considered; it becomes higher as the particle slows down. It is further assumed in calculation that the energy loss occurs only along the pathway, whereby the δ-rays are neglected.

Nevertheless, the LET concept provides a very meaningful overall description of an ionizing radiation. Such knowledge is useful when results are compared for radiations of different types and energies, or for one type of radiation at different energies. In some cases it also facilitates the experimental approach. For example, simulation of the chemical action of cosmic rays in the laboratory requires a source of protons with gigaelectronvolt (1000 megaelectronvolts) energies, since such protons are preα -PARTICLE   FAST ELECTRON

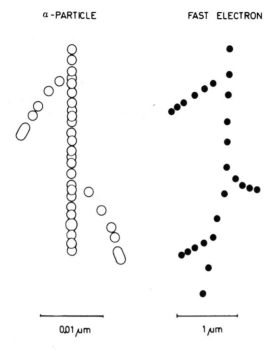

0.01 μm   1 μm

FIGURE 18. Spatial distribution of primary events in water for α-particles (5 megaelectronvolts) and fast electrons (1 megaelectronvolt). Spur lengths are approximately 0.002 micrometer and are not drawn to scale in the case of electrons.

dominant in space. Appropriate machines for this purpose exist, but they are scarce and expensive to operate. Moreover, the fluxes are low and insufficient amounts of radiolytic products are formed for routine analysis. However, an average LET value of cosmic ray proton is 0.2 kiloelectronvolt per micrometer, i.e., similar to that of the γ-rays of radioactive ^{60}Co. Thus, simulation experiments can be satisfactorily performed with radioactive cobalt sources, which are common in research laboratories and convenient for irradiation under a wide range of conditions with respect to temperature, sample volume, and absorbed dose.

Ice

The physical state of an irradiated substance does not affect primary events such as energy transfer, excitation, and ionization. However, it does influence the fate of the primary species formed, as can be seen from the different radiolytic behavior of liquid water, ice, and vapor.

The formation of primary species along the radiation pathways in a rigid structure like ice occurs in the regions of dense energy deposit, i.e., the spurs ("beads") and tracks, in the same way as in liquid water. Most of the electrons ejected by ionization return to their parent or to a neighboring atom that has lost its electron. More energetic electrons continue on their own and ionize or excite further water molecules until their energy has diminished to the point that they are attracted by a parent-like ion, or are retained by a lattice vacancy; in this case, they are called "trapped" electrons (e_{tr}^-). Reactive species such as hydroxyl radicals or hydrogen atoms are also mobile and cause oxidations, reductions, or other chemical changes that are characteristic for liquid water.

A rigid structure, however, strongly limits the mobility of such species, particularly at low temperatures. In consequence, the recombination to water in irradiated ice is enhanced and subsequently the number of reactive chemical species available for reactions is significantly reduced. Accordingly, the radiation chemical yields in ice and frozen aqueous solutions are low in comparison with those in the liquid phase. Only 0.5 water molecule is decomposed per 100 electronvolts of ^{60}Co γ-rays absorbed in ice at 77 degrees kelvin, as compared to 4.5 in liquid water at room temperature (293 degrees kelvin).

In irradiated frozen aqueous solutions essentially the same type of chemical change is observed as at room temperature. This is the case for the oxidation of iron ions or the decomposition of organic compounds like alcohols. However, the corresponding radiation chemical yields are lower by a factor of 10 or more.

Usually the frozen samples are melted prior to analysis. This means that certain reactive species such as e_{tr}^-, which were formed during irradiation and became trapped in the ice structure, become mobile in the course of thawing. These react with neighboring atoms and molecules and, hence, the analysis of an irradiated sample includes not only the products directly produced by radiolysis, but also those from secondary reactions during thawing.

Following melting of irradiated water ice, hydrogen atoms, hydroxyl radicals and e_{tr}^- are no longer found in the sample. However, their presence in the sample before and during the temperature rise can be ascertained by certain techniques. The most important of these is electron paramagnetic resonance (EPR), which is based on magnetic properties associated with the unpaired electron of the free radical. For this purpose, the frozen sample is placed during irradiation in a strong magnetic field and simultaneously subjected to microwaves. The sample selectively absorbs waves which are characteristic for the reactive species. The resulting absorption can be recorded by electronic equipment. It is proportional to the free radical concentration, which changes on warming.

When the water ice is irradiated and measured at 4 degrees kelvin, the hydrogen atoms (H) can be observed by the EPR technique. As the temperature rises, the intensity of the recorded signal decreases and completely disappears at temperatures above 77 degrees kelvin. After thawing the hydrogen is found in molecular form as H_2. Similar observations pertain to the hydroxyl radical. Since the latter is heavier, its mobility is hindered to a greater degree by the rigid structure of ice. This species first disappears at temperatures between 100 and 130 degrees kelvin; subsequently, hydrogen peroxide can be measured in the melted sample and its yield is considerably lower than in a sample irradiated at room temperature.

Water Vapor

The same phenomena of ionization and excitation which take place in the condensed state also occur in irradiated water vapor. The dissociation of water molecules leads not only to the hydrogen atom and hydroxyl radical, but also to molecular hydrogen and atomic oxygen (O), the latter further giving rise to molecular oxygen (O_2). In the vapor phase, the active species diffuse more rapidly and the radiation chemical yield of decomposition may be as high as 8, which is considerably greater than in condensed media. This yield is relatively independent of the LET, i.e., of the type and energy of radiation.

Since the extent of reformation of water from the reaction of the hydrogen atom with the hydroxyl radical is limited, the number of primary species available for reaction is considerably larger and the yields of chemical reactions such as oxidation, reduction, or polymerization are much higher than those in condensed media.

Free Radicals

These consist of atoms or groups of atoms which show increased chemical reactivity owing to the presence of an unpaired electron. Free radicals are produced when a molecule is severed in such a manner that one of the bonding electrons, designated by a dot, remains associated with each fragment:

$$H{:}OH \rightarrow H^\bullet + OH^\bullet$$

The dot representing the unpaired electron is more often used for complex radicals but generally omitted for simple ones such as those produced in water radiolysis.

Free radicals are of primary importance in radiation chemistry. Thousands of these species have been identified despite the fact that most of them exist for only a small fraction of a second. Some have longer lifetimes but usually only under special conditions. At the low temperatures and densities of matter in interstellar space, all free radicals have a significantly prolonged lifetime and play an important role in cosmic chemistry (Chapter 6).

The reactivity of free radicals is reflected in the various ways in which the state of having an unpaired electron is lifted. One of these ways includes electron donation, i.e., reduction, as in the reaction of a hydrated electron with oxygen,

$$e^-_{aq} + O_2 \rightarrow O^-_2 + H_2O$$

and another is electron acceptance as in the oxidation of iron ions,

$$OH + Fe^{2+} \rightarrow OH^- + Fe^{3+}$$

The abstraction of hydrogen represents another free radical reaction which is common for organic molecules. An example is methanol ($CH_3 OH$), which undergoes the reaction:

$$H + CH_3OH \rightarrow H_2 + {}^\bullet CH_2OH \text{ or } OH + CH_3OH \rightarrow H_2O + {}^\bullet CH_2OH$$

The resulting species is an organic free radical with the unpaired electron located on the carbon atom. This radical can disappear by various pathways, one of which is recombination with formation of ethylene glycol, a viscous liquid with a sweet taste used as antifreeze for car motors or as a common starting material in the organic chemical industry.

Free radical reactions explain why pure water does not decompose to an unlimited extent and why a large accumulation of molecular hydrogen and hydrogen peroxide is prevented even in the heavily irradiated water in the core of a nuclear reactor. The recombination reactions of H into H_2, and OH into H_2O_2, are followed by the destruction of these molecules according to

$$H_2 + OH \rightarrow H_2O + H \qquad H_2O_2 + H \rightarrow H_2O + OH$$

In pure water, molecular hydrogen and hydrogen peroxide are both decomposed and formed, and once their equilibrium concentration levels are established they remain practically unchanged.

Trace amounts of impurities in water can be unpleasant if not dangerous. In our context, an example is the presence of dissolved chromium arising from corrosion of the wall material of a nuclear reactor. Even of concentrations of several parts per million, the chromium ions react with primary water radicals and affect the reactions which maintain low equilibrium concentrations of molecular hydrogen and hydrogen peroxide. In consequence, the hydrogen peroxide concentration increases and with it the degree of corrosion. In the routine operation of nuclear reactors which use water as a moderator and coolant, this difficulty is overcome by introducing hydrogen gas into the reactor core. In this way, the above reaction in which hydrogen atoms are produced from molecular hydrogen is enhanced and these, in turn, serve to decompose the detrimental hydrogen peroxide.

Radiation Processing

Chemical modification of materials in industry can often be conveniently performed by radiation processing, although larger doses, of the order of 10 kilograys or more, are generally required. The desired changes are usually produced by the action of radiation-induced free radicals and ions in polymers, which are long chain-like molecules containing repeated combinations of structural units.

In this way, an important degree of quality improvement is achieved in polyethylene, one of the most widely radiation-processed polymers. Polyethylene is a tough, waxy, thermoplastic material which is obtained by polymerization of gaseous ethylene (CH_2CH_2). It is commonly used as an electrical insulator and for many applications calling for a flexible, chemically resistant material. Depending on the manufacturing process, polyethylene softens at temperatures between 70 and 90 degrees Celsius and melts to a viscous liquid at 115 to 125 degrees Celsius. Following treatment by radiation it withstands heating to 250 degrees Celsius without alteration. This improved thermal stability arises from cross-linking, a process in which the free radicals induce a bond to the adjacent polymer chains by lateral links. Material treated in this manner can be used for insulating electric cables designed for high current loads and operating temperatures.

Another application of cross-linked polyethylene involves its "viscoelastic memory". If radiation-processed polyethylene is heated above its melting point, the crystallites melt, but instead of flowing the material exhibits rubber-like elasticity. It can be readily stretched and, after cooling, retains its extended form at room temperature. Now the material can be retreated by heating above its melting point, whereupon the cross-links cause a rubber-like contraction and shrinkage to the original shape. This serves as a basis for the manufacture of "heat-shrinkable" wrapping films and similar products.

The tire industry uses radiation processing of crude rubber, a procedure which is also based on cross-linking and which contributes to the high performance characteristics of vulcanized tires.

Radiation is also useful in developing new polymeric materials such as graft copolymers; these combine the desirable properties of polymeric constituents in the same way that an alloy exhibits the best properties of its metallic components. The mechanism of radiation grafting is based on radiation-induced formation of reactive sites on the main polymer chain; these sites sequentially combine with other reactants, either monomers or polymers, to form a pendant grafted chain. This results in modification of surface properties such as adhesion or wettability without significantly affecting the properties of the bulk material. Radiation grafting has successfully been applied in upgrading textile fibers in order to improve dye fastness or reduce the effect of accumulation of static electric charge.

The bulk properties of less expensive materials such as wood can also be improved by graft copolymerization if the grafted side chains are randomly distributed throughout the object. This is the case for wood-plastic combinations. In the process, a monomer is soaked into the porous wood under pressure and the impregnated material is irradiated. The resulting wood-plastic is a hard material which can exhibit a shiny surface when polished and a remarkable stability in a moist atmosphere.

The improvement of surface coatings is achieved by a procedure called radiation curing. A paste is deposited on the coated surface and subsequently exposed to irradiation. Such processing improves the quality of magnetic tapes for sound and video recording. Radiation is also applied to coatings deposited on paper, metal, or plastic foils.

BIOLOGICAL ACTION

Biological effects of radiation are induced by the absorption of energy in tissues, i.e., in aggregates of similar cells and intercellular material which form the structure of a living body. Most of the radiation energy is absorbed in the water content, which represents about 65 percent by weight, and the final effect is essentially a consequence of reactions of free radicals of water with the chemical constituents of the cell.

Studies of radiation-induced effects at the cellular level are carried out on cells grown *in vitro*. A cell is comparable to a chemical factory in which thousands of coordinated chemical processes take place simultaneously and this complexity renders investigations difficult. Useful information is provided by studies of radiolytic behavior of individual cellular constituents.

The effects of radiation on the human body mainly involve damage to individual cells. Two main categories of harmful effects exist: one of these affects the irradiated person himself (somatic effect) and the other concerns offspring and future generations (hereditary effect). Presently available knowledge is mainly provided by observations of victims of nuclear bombardment of Japanese cities and control of persons who are professionally exposed to radiations, such as personnel in radiological institutions or workers in various fields of nuclear energy. For observation of genetic consequences, a large population and time scale are necessary, such as the statistical survey of population in the area of the nuclear accident at Chernobyl Ukraine.

Biochemical Molecules

Proteins and nucleic acids are the building blocks of living matter and their behavior under the action of radiation is of primary importance for the understanding of radiation damage.

Proteins constitute the principal nitrogeneous organic compounds in living matter and contain about 15 percent nitrogen and 50 percent carbon. They are polymers with molecular masses ranging from 5000 to 6 million atomic mass units. They consist of hundreds or thousands of amino acids joined together by a peptide link (-CO-NH-) to form a chain structure. A protein may contain more than one peptide chain. The three-dimensional arrangement of peptides is largely maintained by hydrogen bonds and is very important in determining properties of the protein. The constituents of a protein comprise about 20 different amino acids and each protein polymer may contain all of these arranged in a variety of sequences. A particular sequence in an individual protein confers specific properties.

Significant information on the radiolysis of proteins in general can be obtained from studies of enzymes, which are an important group of proteins produced by living cells. Enzymes function as highly specific biological catalysts; that is, without undergoing a permanent change themselves, they induce a chemical reaction in the cell more rapidly than it could otherwise occur. Experiments show that an irradiated enzyme loses its enzymatic activity as a result of damage at its active site. However, a radiation dose which completely inactivates an enzyme in aqueous solution may cause only very little loss of activity in the dry enzyme. This is because in the dry form free radicals are less abundant and less mobile than in the solution, with the result that fewer active sites are damaged. It has been found that radiation also damages the specific three-dimensional structure of the protein which is essential to the enzyme's function; the attack of free radicals can disrupt hydrogen bonds or other chemical links, and thus lead to the unfolding of peptide chains and the subsequent loss of enzymatic activity.

Nucleic acids are responsible for storing and transferring hereditary information. They consist of large molecules composed of chains of nucleotides. Each nucleotide contains a nitrogeneous base (purine or pyrimidine), a pentose sugar, and a phosphate group. Nucleic acids are of two types: DNA (deoxyribonucleic acid) which is found in the nuclei of cells and RNA (ribonucleic acid) which is found mainly in the cytoplasm. The molecular structure of RNA is similar to that of DNA except for the type of sugar and one of the bases.

The chemical action of radiation on DNA involves alterations of both the nitrogen bases and the backbone, which consists of two helical chains coiled around the same axis. Rupture of the DNA double strand is a lethal event within the cell. It has been found that a major reaction involves addition of hydroxyl radicals to the double bonds of purines and pyrimidines. The hydroxyl radicals are responsible to an extent of approximately 70 percent for cell mortality, the remaining 30 percent being mainly due to the direct action of radiation on the DNA molecule.

The Cellular Level

Cells are structural units of the living body which usually comprise two main distinct forms of cellular matter: a membrane-bonded body called the nucleus and a surrounding mass known as the cytoplasm. The nucleus organizes the synthesis of enzymes and controls the characteristics of the cells' progeny. It contains chromosomes, which are simple proteins associated with coils of heredity substances, i.e., DNA. The "work" within the cell is carried out in the cytoplasm, where enzymes control the cell's metabolism and the manufacture of its constituents.

Studies are frequently carried out with cells which are grown *in vitro*. A direct bombardment of the cell nucleus is much more efficient than the irradiation of cytoplasm. The microbeam used for this purpose is a specially fine and sharply focused beam of ionizing radiation. In some experiments a radioactive isotope is incorporated into the nucleus to provide *in situ* irradiation. For this purpose the hydrogen in DNA can be replaced with tritium, the radioactive isotope of hydrogen. Its low-energy β-rays remain mostly localized at the site of decay.

It is interesting to examine both normal and tumorogenic human and animal cells *in vitro* following irradiation in the presence or absence of certain drugs which act as modifiers of radiation effects. These drugs may increase the effect, as in the case of sensitizers or minimize it, whence they are called protectors. The chemicals do not interfere with fast, early processes such as ionizations, excitations, and free radical formation. However, they can interfere with the reactions of free radicals once they are formed and thus influence the extent and nature of radiation damage at the cellular level. In general, oxidants such as oxygen enhance or fix damage and act as radiosensitizers producing a higher extent of damage per unit of absorbed dose. On the other hand, reducing agents such as vitamin C have the opposite effect and tend to repair the damage, hence providing radioprotection. Studies of the chemistry of potential radiation modifiers are of great interest in cancer therapy. The radiotherapist attempts to maximize the tumor dose while still maintaining the dose received by surrounding healthy tissues at suitable tolerance levels.

The Human Body

As we have seen, radiation acts on the human body through chemical changes induced at the cellular level. These interactions may affect the individual cells in a number of ways, of which examples are premature cell death, prevention of cell division, or a permanent or genetic modification which is passed on to daughter cells.

Fortunately, not all damage to individual cells is permanent; the body's repair mechanisms contend with the situation. The lower the dose and the dose-rate, the higher is the efficiency of repair. To produce an acute injury, the radiation factors must reach certain levels. However, these levels appear to be very low when we consider conditions leading to the induction of cancer or of genetic damage for which, most likely, the smallest dose presents a risk. Because of this, no level of exposure to radiation can be considered as "safe".

In man, the harmful consequences of irradiation become noticeable at different times following exposure. It may be a matter of hours or days for early effects that are mainly due to damage to bone marrow, the gastrointestinal tract, or neuromuscular regions. It may take years or decades for the late effects which appear as leukemia and other forms of cancer. In the event of hereditary effects involving damage to chromosomes and genes, the effects will first become apparent in the following generations.

Of the various factors which determine the type and extent of radiation damage, some depend on the individual and his health, age, and sex. Others involve the irradiation conditions, and whether the whole body or part of it is exposed. The dose rate, and above all the absorbed dose, are particularly important.

At very high dose levels of about 100 sieverts received by the whole body during periods up to several hours, radiation is rapidly lethal, mainly because of severe damage to the central nervous system. At doses of 50 to 10 sieverts, there is partial damage to the central nervous system together with destruction of cells lining the intestine. The resulting gastrointestinal disturbances are followed by severe bacterial invasion and death within 1 or 2 weeks. However, such high doses are exceptional and occur generally only in events such as major accidents in the field of nuclear energy or in the use of nuclear explosives. Very little data are available for human exposure in this dose range and experiments on animals are the main source of information.

A whole body irradiation during several hours at doses above 1 sievert gives rise to nausea and vomiting. This disease is known as "radiation sickness" and occurs a few hours after exposure as a result of damage to cells of the intestinal wall. The mortality risk is low at doses below about 1.5 sieverts, but at 8 sieverts, the prognosis is poor. When doses up to 10 sieverts are fatal, death is usually due to secondary infections because of depletion of the white blood cells, which normally provide protection against infection. The chances of survival in such cases can be increased by special medical treatment, which includes isolation in a sterile environment and the stimulation of leukocyte production.

There is no well-defined radiation dose for which death is certain. It is considered that because of damage to the bone marrow, doses between 3 and 4 sieverts have lethal effects within a month in about half of the exposed people and in the absence of the more highly specialized forms of medical treatment. This estimate is based on the mortality rate observed in various accidents involving whole body exposures.

Various parts of the human body respond very differently to radiation. Most sensitive are the bone marrow and the circulatory system which are affected by as little as 0.5 sievert. However, regeneration may be efficient if only part of the body is irradiated, and surviving bone marrow can replace the damaged portion. The reproductive organs are also particularly sensitive and single doses above 2 sieverts received by the testes or 3 sieverts by the ovaries can cause permanent sterility. The eye is also a very vulnerable organ and serious loss of vision due to radiation-induced opacity of the lens may be caused by single doses between

2 and 5 sieverts, or even by much lower doses (down to 0.5 sievert) if accumulated over a period of years, as in occupational exposure. It is important to note, however, that these and other organs receive considerably larger doses in radiotherapy. In this case, such doses are taken into consideration only when the benefit is greater than the risk. Such treatments are then carried out successively and use, for example, about 6 sieverts for the bone marrow or ovary, and 14 sieverts for the eye lens.

Children are in every respect much more vulnerable to radiation than are adults; the younger the child is the greater the damage may be. Unborn children are particularly susceptible to brain damage if their mothers are irradiated between the 8th and 15th week of pregnancy, since this is the period when the cortex of the brain is formed. Such radiation-induced changes can cause severe mental retardation.

Cancer is the most important consequence of lower doses of radiation. The latter can damage the control system of a single cell and cause it to divide more rapidly than usual. The induced defect is transmitted to the daughter cells and the population of abnormal cells increases with respect to that of normal cells in a body organ. A long and variable latent period may exist between exposure to radiation and the appearance of cancer, which may become manifest after 5 to 30 years or more. The leukemias occur first, in most cases after about 6 to 7 years, although an earlier appearance is possible. The risk of all other forms of cancer becomes significant 25 years after exposure. However, solid tumors may appear already after 10 years and the risk increases with time.

Genetic mutations are caused by damage to the hereditary material of the cell, i.e., in the chromosomes, which are thread-like constituents of nucleus. They contain genes, the carriers of information which determines the characteristics of daughter cells. Here, radiation-induced damage causes chromosome aberration, involving changes in their number or structure and mutations of the genes themselves. These mutations are of two types: dominant, which concerns the children of persons already affected by radiation, and recessive, which may be dormant for many generations and first becomes manifest in a child whose parents both possess the mutant gene. Many diseases are associated with recessive genes. Since mutant genes are generally recessive, it is assumed that virtually all mutations are harmful. Ionizing radiation increases the rate of mutation and, in the prevailing circumstances, it is to be feared that an increased population of genetically abnormal individuals could appear in future generations.

Radiation exposure and radiation protection of population are considered in Chapter 8.

Risks

Irradiation may be compared to gunfire; for a living target there is no safety level for a striking bullet, but not every bullet that does hit will kill. An individual exposed to radiation is simply subjected to a greater risk than if he had not been irradiated. And, obviously, the risk increases with the size of the dose.

The word risk is used to state the probability of severe harm due to radiation exposure. When taking into account retarded effects of radiation, such as cancer and genetic changes, a reliable assessment of risk for exposed persons is of paramount importance. Estimates of nominal risks of cancer and genetic defects caused by ionizing radiation are undergoing continuous review by various scientific bodies, such as the UNSCEAR (United Nations Scientific Committee on the Effects of Atomic Radiations). Proposed figures are submitted to changes as new data become available from observations of larger populations, or from laboratory studies.

In the mid-1980s, the estimate of an average risk of cancer per millisievert per person was about 12.5 in a typical population of a million individuals of both sexes, of all ages. This means that if each of 1 million persons receives a dose of 1 millisievert, the number of fatal cancers would be about 12.5 and would appear over a period of about a decade. This value is considered by many scientists to be an underestimate. According to a survey conducted in 1980 by the U.S. National Academy of Sciences and the Committee of Biological Effects of Ionizing Radiation (BEIR), the risk may be as high as four times greater, i.e., up to 50 persons per million per millisievert.

The risk of hereditary effects is much more difficult to estimate than that of cancer because of limited understanding of the radiation-induced genetic changes and a lack of available data. A figure proposed in the mid-1980s by the ICRP (International Commission on Radiological Protection) is about eight significant cases per millisievert per person in a typical population of 1 million. It represents the total genetic risk in all generations averaged over both sexes and all ages.

The above figures are indicative for assessments of harmful consequences of population irradiation in case of accidents, whether of a more limited character like radioactivity released in a local nuclear plant, or accidents with global consequences in the case of Chernobyl (Chapter 9). A major obstacle in such evaluations is often the difficulty of acquiring accurate information on the radiation doses involved.

A detailed revision of the generally accepted dosimetric data on the atomic bomb victims in Japan was published in 1986, and revealed that the real absorbed doses received by the population of Hiroshima and Nagasaki must have been considerably lower than was previously assumed. This implies that low radiation doses are much more dangerous than hitherto supposed and that the risk estimates given above should be extensively revised.

Risk implies uncertainty, and risk assessments are not easily visualized unless they can be compared with familiar concepts. For example, a rough estimate is that about 2000 deaths from cancers will be due to other causes in the same typical population of 1 million within a period of 1 year.

Radiation Sterilization

Lethal effects of ionizing radiation are used on an industrial scale for sterilization, i.e., for destroying contaminating microorganisms or reducing their number to an acceptable level (Chapter 8).

An important feature of radiation sterilization is that the necessary doses do not significantly increase the temperature. This permits sterilization of thermolabile plastics which would not withstand the temperatures of heat sterilization procedures. Very often, "cold" sterilization is the only possible method of sterilizing biological tissues and various preparations of biological origin. It appears also to be the only means of sterilizing a number of heat-sensitive pharmaceuticals (powders, ointments) and certain cosmetic raw materials or finished products.

A further advantage of radiation sterilization is high penetrating power of radiation, enabling access to all parts of the object and the wrapping material. Thus, the items can be prepacked in hermetically sealed packages which are "impermeable" to microorganisms. In the medical profession, this is particularly useful in the marketing of sterilized sutures and disposable medical supplies such as syringes, catheters, and hospital clothing. At the present time, most of the 150 irradiators using radioactive cobalt are employed for this purpose throughout the world.

Since radiation kills microorganisms, it can be used for food conservation by reducing the levels of disease- and mold-producing microbes. This is especially desirable for the storage and transport of highly perishable foods, such as fish or shrimps. It should be noted, however, that irradiation does not remove toxins created by bacteria during the period of contamination.

Table 9

**DOSE-RANGES IN KILOGRAYS REQUIRED FOR
SPECIFIC RADIATION TREATMENTS**

Sterilization of insects and parasites	0.03—0.2
Killing of insects and parasites	0.05—5
Million-fold reduction in the number of bacteria, moulds and fungi	1—10
Million-fold reduction in the number of viruses	10—40
Sterilization of food	20—45

Note: 1 kilogray = 1,000 joules/kilogram = 100,000 rads.

Radiation has been found to be particularly effective in de-stroying insects and their eggs in flour, or reducing the levels of Salmonella or Lactobacilli in poultry eliminating the microorgan-isms in precooked meats. As a general rule, it may be stated that the smaller and simpler the organism is, the higher will be the dose necessary for lethal action (Table 9).

It should be pointed out that there is no induced radioactivity in the commodities sterilized by radiations; the γ rays of ^{60}Co cannot cause nuclear reactions or produce radionuclides, and this is also true for electrons if their energy is maintained below 10 megaelectronvolts. Nonetheless, radiation also induces chemical changes which accumulate in irradiated food. The irradiation of meat up to 10 kilograys causes partial degradation of proteins and lipids with some formation of volatile compounds; at larger doses, an unpleasant odor and flavor can develop as a result of the accumulation of radiolytic products. Knowledge of the chemical composition and amounts of these products is essential in main-taining the wholesomeness of irradiated food and further studies are necessary before wider public approval can be achieved.

Radiation sterilization of spices is particularly useful in the canned food industry. Irradiation of the solid material efficiently eliminates the putrefactive microorganisms without altering the flavor or producing harmful radiolytic products.

Relatively low doses of 10 to 100 grays are used for extending the shelf life and preventing the sprouting of potatoes or onions during periods of up to a year. With a dose of 2 kilograys, strawberries can be stored without refrigeration for a week. The extent of radiation-induced chemical changes is negligible, and the flavor and physical characteristics are unchanged.

Sterile irradiated food is used on some occasions when the benefit is greater than the risk, as in the case of the Russian and

American crews on the Apollo-Soyuz space mission in 1975. In order to reduce risks of infections contracted from nourishment, patients who have received a bone marrow transplant, or AIDS patients, are sometimes fed with irradiated food. The use of food preserved by radiation is a routine practice for feeding laboratory animals that are bred and maintained under sterile conditions.

LARGE SOURCES

Ionizing radiation is a specific source of energy for promoting chemical changes, and even at powers in the range 1.5 to 60 kilowatts as for ^{60}Co units, or 25 to 200 kilowatts as with accelerator electrons, it is a powerful tool in industry. In comparison, 1 kilowatt of electric power is just enough to run a toaster or a laundry iron, whereas an installation of radioactive cobalt at a power of 1 kilowatt can process thousands of tons of material annually.

^{60}Co is prepared by irradiating pellets or small disks of metallic cobalt in nuclear reactors. Following irradiation, these are assembled into radiation sources of desired size and intensity, usually in the form of a plaque. An industrial unit with radioactive cobalt often contains about 37,000 terabecquerels (1 million curies). When not in use, the plaque is usually stored in a pool of water 5 meters deep to provide radiation protection for the personnel and absorb the heat released from radioactive decay. In routine use, the plaque is placed in a concrete vault containing the items to be processed (Figure 19).

The γ-rays of ^{60}Co penetrate to a depth of several decimeters in moderately dense material. Large packages can thus be irradiated at uniform doses such as are needed for sterilization. A 1-megacurie (1 million curies) source of ^{60}Co has a power of 15 kilowatts and provides typical dose rates between 1 and 10 kilograys per hour; in a restricted volume, a maximum of 100 kilograys per hour can be achieved. The dose rates decrease by about 1 percent per month because of the 5.3 year half-life of ^{60}Co.

An electron accelerator consists of an accelerator tube and a high voltage supply. The energies of electrons from industrial accelerators vary between 0.2 and 10 megaelectronvolts. Electrons at low energies have a penetration power of a few centimeters at most, and are suitable for surface curing, or for other chemical applications such as cross-linking and grafting. The upper limit of 10 megaelectronvolts is deliberately chosen to avoid the formation of radionuclides in irradiated material. Accelerators provide considerably larger dose rates than isotopic sources, and can deliver up to 100 kilograys per second.

The safety of personnel and of the nearby population, particularly with respect to minimization of the risk of accidental radiation exposure, is a major problem in the design of an industrial radiation processing plant. In addition to the shielding and interlock system, efficient ventilation is also required since ozone and noxious nitrogen oxides are produced in very high radiation chemical yields which may amount to as many as 10 molecules per 100 electronvolts absorbed in air (Figure 19).

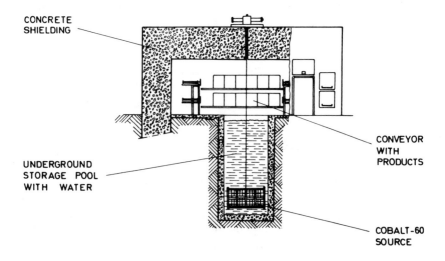

FIGURE 19. Elevation view of a typical ^{60}Co irradiation facility for sterilization.

Further Reading

Radiation, Doses Effects, Risks, United Nations Environment Programme, UNEP, 1985. [A brief survey of facts on natural and artificial sources of radiation, their effects on man, and the subsequent risks. Suitable for the general public, it is well written and illustrated.]

Hughes, G., *Radiation Chemistry,* Oxford Chemistry Series, Clarendon Press, Oxford, 1973. [A clear presentation of principles of the chemical action of ionizing radiation.]

Swallow, A.J., *Radiation Chemistry, an Introduction,* Longman, London, 1973. [Gives a well-balanced picture of the subject as a whole.]

Spinks, J.W.T. and Wood, R.J., *An Introduction to Radiation Chemistry,* 2nd ed., John Wiley & Sons, New York, 1976. [Basic principles of radiation interaction with matter, the physical stage, sources, and dosimetry are presented clearly. An excellent, very detailed introduction to various aspects of radiation chemistry and its applications. It has been written mainly for those specializing in radiation chemistry, but should be very useful in related fields such as nuclear technology or radiobiology. An extensive survey of important publications in radiation chemistry is given up to 1975.]

Gaughran, E.R.L. and Goudie, A.J.,Eds., *Technical Developments and Prospects of Sterilization by Ionizing Radiation,* Vol. 1 and 2, Multiscience Publishing, Montreal, 1974 and 1978. [Written by professionals for professionals, it can also be used by those who have a general education in natural sciences and an interest in isotopic irradiators, accelerators, basic physico-chemical aspects of radiation (Vol. 1), and the biological application of sterilization (Vol. 2).]

Martin, A. and Harbisan, S.A., *An Introduction to Radiation Protection,* 2nd ed., Chapman and Hall, London, 1979. [A comprehensive account of radiation hazards and their control, this book assumes no previous knowledge of the subject. Of interest to general readers also because of an excellent presentation of basic principles of the structure of matter, radioactivity and radiation, as well as of detection and dosimetry of radiation.]

Farhatazis, and Rodgers, M.A.J., Eds., *Radiation Chemistry and Applications,* Verlag Chemie, Weinheim, 1987. [An up-to-date, comprehensive presentation of fundamentals of radiation chemistry and of its main applications. Prepared by 22 experts, it is a book that should be consulted.]

Sonntag, C., *The Chemical Basis of Radiation Biology,* Taylor & Francis, London, 1987. [Clearly written, detailed presentation for those who intend to undertake the research in the field of biological action of radiation. An extensive survey of literature is given up to 1987. Despite the scientific style of presentation, the book can be used by nonprofessionals who have a more profound interest in this branch of radiation science.]

The radioactive isotopes found in nature (about 80) are largely exceeded in number by those made by man (over 2500). Artificial radionuclides are obtained usually in nuclear reactions induced by the bombardment of a target nuclide with charged particles or with neutrons. This chapter deals with various chemical aspects of nuclear processes.

In many nuclear reactions a significant amount of energy is released by the transformation of mass into energy. This conversion is particularly important in the fission reaction, in which a target nucleus splits into two fragments. The fission of uranium induced by neutrons is the basic reaction for nuclear energy.

In radioactive decays and in nuclear reactions the nuclides are formed with kinetic energies which may be higher than the energy of chemical bonds. These species are known as "hot atoms". They permit the study of chemical reactions of energy-rich entities. Hot atoms intervene frequently in geochemistry and cosmochemistry and are convenient for the fast synthesis of molecules labeled with short-lived radionuclides.

A unique feature of radioactive decay is the transmutation effect by which the chemical identity of the daughter nucleus differs from that of the parent. Such effects can be used for the decay-induced synthesis of new molecules; they may have played a role in the formation of precursors of biomolecules on the early Earth and are likely still important for cosmic chemistry.

The chemical environment may influence nuclear properties and a very weak change of half-life is observed in a few instances.

4

Chemical Aspects of Nuclear Processes

NUCLEAR TRANSFORMATIONS

An atomic nucleus is characterized and defined by the number of its nucleons, consisting of neutrons and protons. When the structure of the nucleus is altered either by a gain or loss of some of the nucleons or by a change in the neutron/proton ratio, the process is called a nuclear transformation. It may occur spontaneously in nature, as in radioactive decays observed on the Earth or in more complex processes which take place in stars. Nuclear transformations are realized artificially with devices such as reactors and accelerators, or in nuclear explosions.

Nuclear Reactions

For performing a nuclear transformation, scientists have a rather large choice of nuclear reactions. These are carried out by bombarding a target nucleus with an appropriate projectile such as a proton, a neutron, some other nucleus, or an energetic electromagnetic radiation. In the course of the nuclear reaction a new nucleus is formed, which has a neutron/proton ratio different from that of the target and may be either stable or radioactive.

As an example we consider a target consisting of nitrogen. The nucleus of its most abundant isotope has 7 protons and 7 neutrons, i.e., 14 nucleons, and is represented by the symbol $^{14}_{7}N$. If one of the protons is replaced by a neutron, a new nucleus is formed which still contains 14 nucleons, but now composed of 8 neutrons and 6 protons. This nucleus is that of a radioactive isotope of the element carbon, $^{14}_{6}C$. In order to perform this reaction, which is a genuine transmutation of one chemical element into another, a neutron (the projectile) has to be forced into the original nitrogen nucleus (the target) and, at the same time, a proton must be expelled. The representation of this nuclear reaction is

$$\text{nitrogen-}14 + \text{neutron} \rightarrow \text{carbon-}14 + \text{proton}$$

In the same manner as for chemical reactions, the reacting species is denoted on the left side and the products on the right. The shorthand representation is $^{14}_{7}N$ (n, p) $^{14}_{6}C$.

The total number of nucleons, as well as that of the individual protons and neutrons, is always preserved in a nuclear reaction. Quite often the total mass of the products of the nuclear reaction is smaller than that of the sum of target nucleus and projectile. In this case, the missing mass is converted into energy according to Einstein's relation, $E = mc^2$. This energy appears as kinetic energy of the emitted particle. In other instances, the products of the reaction are heavier than the total initial mass of the reacting species. Here, energy must be supplied to induce the reaction. This is usually achieved by conveying sufficient kinetic energy to the projectile, whereby part of the energy is converted into mass. Energy can also be furnished by heating the reactants, but in nuclear chemistry the amounts of heat required are enormous. In the laboratory, only the fusion of hydrogen to deuterium can be realized by supplying heat, and then only under very special conditions (Chapter 10). In the Sun and stars such a reaction is quite common.

The occurrence of a nuclear reaction is a matter of chance. When a proton strikes a target atom, it has a much higher probability of encountering the expanded core of the electrons rather than the tiny nucleus inside. The probability of the reaction is expressed in terms of cross-section, which can be visualized as an imaginary disk covering the nucleus. The reaction is considered to occur following an impact of the projectile with the surface of the disk. The larger the "surface", the higher the probability of the reaction.

The radius of a heavier nucleus is of the order of 10^{-14} meter, and consequently the geometrical cross-sectional area is about 10^{-28} square meter. The term "barn", corresponding to this latter value has been selected as the unit of cross section. Cross section values vary widely (Table 10). For many reactions, they are of the order of a few barns.

Table 10

CROSS SECTIONS OF NUCLEAR REACTIONS

Target	Projectile	Energy	Nuclear reaction	Cross section (barns)
^{16}O	Neutron	0.025 eV	$^{16}O(n,\gamma)^{17}O$	10^{-4}
^{16}O	Neutron	14 MeV	$^{16}O(n,p)^{16}N$	0.10
^{113}Cd	Neutron	0.025 eV	$^{113}Cd(n,\gamma)$ ^{114}Cd	20,000
^{27}Al	Alpha	5 MeV	$^{27}Al(\alpha,n)^{30}P$	0.02

When cross sections are well below one barn the reaction probability is low. The yields of reactions used for the production of elements with atomic numbers higher than 106 correspond to cross sections of 10^{-9} barn or less. Consequently, only a few atoms are obtained in an experiment.

Other nuclear reactions involve cross sections of hundreds of barns, such as in the neutron-induced fission of uranium, or even several thousand barns, as in the capture of a neutron by cadmium. Uranium and other materials used in the construction of nuclear reactors must satisfy criteria of "nuclear purity", which necessitates a very low content of cadmium or other elements having large cross sections for neutron capture.

In addition to the target-projectile relationship, the cross section of a reaction depends on the energy of the bombarding particle. With the exception of neutrons, the projectiles bear a positive charge and are subject to a repulsive force when they approach the positively charged target nucleus. The intensity of repulsion increases with the charges of both the target and the projectile. For capture probability to be high, charged particles must have energies equivalent to or greater than the repulsion energy. At least 700 megaelectronvolts are needed in order to promote a reaction between two uranium nuclei.

When particles with moderate energies, i.e. of a few megaelectronvolts, strike a nucleus they eject a single nucleon or an alpha-particle from the target. With increasing energy of the projectile, the bombarded nucleus releases more and more nucleons and is finally smashed completely when the energy of the projectile is in the 100-megaelectronvolt range.

Since they are uncharged, neutrons easily penetrate into all nuclei. The most common reaction involves capture of a neutron and release of energy in the form of a gamma-ray. The shorthand notation for this reaction is (n, γ). The cross section of a (n, γ) reaction increases when the energy of the neutrons decreases and attains a maximum for "thermal" neutrons, whose energy distribution is approximately the same as that of gas molecules at ordinary temperatures. The average energy of thermal neutrons is 0.025 electronvolt and their velocity 2200 meters per second. Thermal neutrons are very important in the operation of nuclear reactors and for the preparation of artificial radioelements.

In the same way as for charged particles, neutrons with energies of a few megaelectronvolts cause ejection of one or several nucleons ([n, p] or [n, 2n] reaction) or of an α -particle (n, α reaction) from the target nucleus. The cross section of these reactions increases with the kinetic energy of the projectile.

In addition to capture by a target nucleus, charged particles can also be diverted from their pathway when they pass the nucleus without producing a nuclear transformation. Such "scattering" is due to the electric repulsion between the particle and the nucleus. It can occur at all energy levels. During this process, which is called elastic scattering, the particle loses part of its kinetic energy. The remaining energy depends on the mass of the elements in the target surface. For analytical purposes, one uses protons of a few hundred kiloelectronvolts or α-particles of a few megaelectronvolts. With these energies, the composition of the first micrometer of a surface can be determined accurately. This technique finds applications in the fields of semiconductors, thin films, and corrosion.

A spectacular use of nuclear backscattering was made in 1967, when the first elemental analysis of lunar soil was performed by Surveyor 5, the first spaceprobe to make a soft landing on the Moon. In this experiment, the 5-megaelectronvolt α-particles from the synthetic radionuclide ^{242}Cm were used to irradiate the lunar surface. The energy of the scattered α-particles was employed in determining carbon, oxygen, sodium, magnesium, aluminum, silicon, and some heavier elements. The results of the analysis agreed satisfactorily with a later analysis of Moon rocks which were brought back to Earth. The surface of other planetary bodies could be analyzed in the same way.

Nuclear Fission

One nuclear reaction which is very peculiar in outcome and dramatic in historical consequences is the fission process. In this, a nucleus splits into two parts and simultaneously releases several neutrons and an enormous amount of energy. Fission may be spontaneous or it can be induced by the impact of a neutron, an energetic charged particle, or a photon.

Only the heaviest elements decay by spontaneous fission. The corresponding half-lives range from 10^{21} years for ^{232}Th and 10^{16} years for ^{238}U down to 2 months for ^{256}Cf, and 2.6 hours for ^{256}Fm, which has an atomic number Z equal to 100. With a further increase in Z, the half-life for spontaneous fission diminishes rapidly and becomes of the order of seconds or less. This trend may well be a limiting factor for the synthesis of further transuranium elements.

The fission process can be conveniently visualized with the aid of a model which represents a nucleus by a liquid drop. In the droplike nucleus two opposing forces are exerted: the disruptive repulsion between the positively charged protons and the force which attracts adjacent nucleons. Nucleons at the surface are not completely surrounded by other nucleons and do not experience the nuclear force as effectively as those located deeper within the nucleus. The total nuclear force acts like the surface tension of a liquid, tending to force the drop into a shape having the lowest surface area, namely, that of a sphere.

Many nuclear phenomena can be rationalized by this model. Upon capture of a neutron or a charged particle, the nucleus becomes excited and behaves like a heated liquid drop, leading to "evaporation" of nucleons. The excess energy can also cause distortion of the spherical shape, owing to movements of the drop's constituents. The drop oscillates between the spherical and nonspherical shapes. If the excitation energy is sufficiently high, the drop can acquire such a distorted shape that the surface energy is unable to restore the spherical configuration. Then the nucleus separates into two parts, undergoing fission (Figure 20).

FIGURE 20. Successive shapes of a nucleus undergoing fission according to the liquid drop model.

Most nuclei with masses higher than 100 atomic mass units can undergo fission following bombardment with accelerated charged particles, high energy γ-rays, and, very importantly, with thermal neutrons.

Three nuclides which readily undergo fission with thermal neutrons are ^{235}U, ^{233}U, and ^{239}Pu (Table 11).

Table 11

CROSS SECTIONS OF THE FISSION REACTION INDUCED BY THERMAL (0.025 ELECTRONVOLT) AND FAST (> 1 MEGAELECTRONVOLT) NEUTRONS

Fission cross section (barns)

Nuclide	Thermal neutrons	Fast neutrons
^{233}U	524	220
^{235}U	582	1.5
^{239}Pu	754	1.8

The first of these occurs in nature, and the other two are made artificially: ^{233}U from natural thorium and ^{239}Pu from ^{238}U (Chapter 9).

In the fission process, the nucleus splits into a heavy fragment with mass number about 140 and a complementary light fragment of mass about 90. The probability of separation into two equal masses is very low. The energy released in the fission of one nucleus is of the order of 200 megaelectronvolts and it appears mainly as the kinetic energy of the two fragments. This energy is converted into heat, and this process is used in power plants.

The energy output of fission greatly exceeds the few megaelectronvolts released in most other nuclear reactions, and is incomparably higher than the heat liberated in a chemical reaction between two molecules. To illustrate this on a macroscopic scale, we may recall that a 1-gigawatt nuclear power plant consumes 1 kilogram of ^{235}U per day.

The nuclides produced in fission are radioactive and decay with emission of β-particles and γ-rays, producing other radionuclides. All radioactive elements thus formed are called fission products. Their number is about 200 and their half-lives range from a fraction of a second to thousands of years. The long-lived fission products represent a large part of the radioactive wastes from nuclear power plants.

Every time a nucleus splits into two parts, a few neutrons are released, and these in turn can trigger the fission of further nuclei in a chain reaction. However, the neutrons released in fission have a broad energy spectrum, with energies up to 10 megaelectron-volts or more, and in order to propagate the chain neutrons must be slowed down to thermal energies. This can be accomplished with a moderator, a substance made of light nuclei such as water or graphite. A neutron which hits a low-mass nucleus loses part of its kinetic energy and is progressively reduced to 0.025 electron-volt by repetitive collisions (Chapter 9).

THE MACHINERY

In order to produce a nuclear reaction, the force which acts between the nucleons must be overcome. This requires a considerable amount of energy, at least in comparison with the energy needed for a chemical reaction.

Fast-moving particles constitute an adequate means for producing nuclear reactions. These particles are abundant in cosmic space and in the terrestrial atmosphere. Being emitted by the Sun or produced in remote parts of the galaxy, they frequently have energies which greatly exceed that of nuclear binding. Outer space is truly a convenient laboratory for many nuclear reactions.

In the laboratory, the most common device for the preparation of radioisotopes with accelerated particles is the cyclotron. This is a circular type of accelerator in which the ions describe a spiral trajectory and attain energies up to 100 megaelectronvolts. Current intensities are typically of the order of a milliampere, which corresponds to a flow of 6×10^{15} particles per second. Machines of more compact form, called "baby cyclotrons" are increasingly being used in hospitals for medical applications of short lived nuclides.

Neutrons are particularly suited for inducing nuclear reactions and for preparing radioisotopes. They are produced in nuclear reactors (Chapter 9) in amounts expressed by the neutron flux, which is defined as the number of these particles passing per second through a surface of 1 square centimeter. Neutron fluxes range from 10^9 to 10^{13} per second per square centimeter, depending on the type of the reactor.

Reactors are less versatile than accelerators, which can be readily put into operation and shut down, and which provide various types of particles at controlled energy. For large-scale production of radioelements, on the other hand, the reactor has no equivalent.

Before the discovery of fission in 1939, small neutron fluxes were obtained in nuclear reactions produced by causing α-particles of natural elements to strike the nucleus of a light element such as beryllium. The neutron output obtained in this way is about a million times lower than that of the smallest reactor.

Nevertheless, these sources made an important contribution to the discovery of fission and chain reactions, and they have also been used for making artificial radioisotopes.

Although the original sources used ^{226}Ra or ^{210}Po, the present ones consist of a mixture of beryllium and an artificial α-emitter such as ^{241}Am, which has a very convenient half-life of 450 years and requires less shielding than radium. The output of a source loaded with 10 curies of americium is about 10^7 neutrons per second. Owing to their small size (10 centimeters), these neutron sources are easily transported and can be employed for various analytical applications in the field. They are also very useful for teaching purposes if a nuclear reactor is unavailable.

Recently, even smaller neutron sources have been developed, which utilize microgram or milligram amounts of ^{252}Cf. This heavy nuclide decays partly by spontaneous fission; each time a californium nucleus decays, an average of three neutrons are emitted. Despite the short half-life of 2 years and, for the time being, the high cost of this radioelement, californium neutron sources appear very promising. A 1-milligram source emits over 10^9 neutrons per second. Numerous applications of such a device can be easily imagined in many fields of science, technology, and medicine.

For the most part, nuclear reactors provide neutrons with rather small kinetic energies, i.e., down to the level of 0.025 electronvolt which is typical of thermal neutrons. If more energetic neutrons are required, reactions initiated by charged particles from accelerators must be considered. A common procedure is the bombardment of a target containing tritium, i.e., radioactive hydrogen, with accelerated nuclei of heavy hydrogen (deuterons). In this case, neutrons are emitted with a kinetic energy of 14 megaelectronvolts and a yield which culminates at the rather low accelerating potential of 150,000 volts. Heavy shielding is required in order to absorb the fast neutrons, which otherwise would travel over hundreds of meters. Fast neutrons have many applications, in particular for the analysis of traces of oxygen in solid materials such as metals and semiconductors.

Accelerator-produced neutrons are also required when complicated measurements must be performed during irradiation. The access to irradiated samples in a nuclear reactor is frequently difficult. Although neutron fluxes from charged-particle accelerators are relatively small in comparison with those of nuclear reactors, their energy can be chosen by appropriate selection of the target material.

γ-Rays are useful in nuclear reaction studies only if their energy exceeds tens of megaelectronvolts. Radioelements are not suitable for this purpose. To obtain high-energy γ-radiations, electrons must be accelerated to several tens of megaelectronvolts in a linear or circular accelerator, whereupon they are allowed to strike a heavy target. In this way the full kinetic energy of the electrons is converted into electromagnetic radiation. Since the latter originates from the slowing down of the electrons, it is termed "bremsstrahlung" (Chapter 3).

HOT ATOM CHEMISTRY

Nuclear Recoil

The discovery of artificial radioactivity stimulated the search for convenient techniques of radioisotope production. Leo Szilard and T. A. Chalmers made an important observation when they irradiated ethyl iodide, an organic liquid, with neutrons. In this compound, the iodine atom is strongly bound to a carbon atom. The stable nucleus of ^{127}I readily captures a neutron and is transformed into radioactive ^{128}I. It was found that, following irradiation, about 50 percent of the radioactive iodine was no longer bound in ethyl iodide and could be easily separated by extraction with water. Thus, chemical bonds are severed in the course of neutron capture, and, in particular, the radioactive atoms can be separated from the stable isotopes. For the first time it became possible to separate isotopes in a very simple and efficacious way. Although the radioactive atoms were not entirely free from small amounts of stable iodine, the degree of separation was quite remarkable. Discovered in 1934, this process is known as a "Szilard-Chalmers" reaction. For a long time this technique was utilized in the preparation of radioisotopes having a high specific activity.

An insight into the Szilard-Chalmers reaction is of interest in understanding the chemistry connected with nuclear processes. Following neutron capture, ^{128}I is formed in an excited state which rapidly decays (within less than a millionth of a second) to the ground state by emission of γ-rays. The golden rule of conservation of energy and momentum in any transformation also holds at the atomic and nuclear levels. The γ-radiation carries a momentum to which the radioactive iodine atom must oppose an identical momentum. In this way the iodine acquires sufficient kinetic energy to break the chemical bond with the carbon atom. It is customary to denote the atom formed in a nuclear reaction as a "recoil" atom. The recoil energy involved in the neutron capture by an iodine atom is about 100 electronvolts, which is far in excess of the 2 electronvolts corresponding to the energy of the carbon-iodine chemical bond. The radioactive atom is ejected from its host molecule.

The recoil phenomenon accompanies all nuclear reactions and is also effective in radioactive decays. Recoil energy depends on the mass of the emitted particles and the kinetic energy of the bombarding projectile. The emission of a heavy α-particle by a nuclide such as ^{238}U in the disintegration

$$\text{uranium-238} \rightarrow \text{α-particle} + \text{thorium-234}$$

may be compared to a howitzer shooting a shell. The velocity of the α-particle amounts to 10,000 kilometers per second while the nascent thorium atom recoils in the opposite direction at a speed of 100 kilometers per second and with an energy of 100,000 electronvolts.

The fast-moving recoil atoms are commonly termed "hot atoms". The reference to temperature arises from the equivalence of mechanical and thermal energies. At room temperature, the most probable energy of an atom is 0.025 electronvolt. Hence, the thermal equivalent to hot atoms is of the order of millions of degrees. It would be impossible to achieve the same energy by heating, inasmuch as 1 electronvolt is equivalent to a temperature of several thousand degrees. Nuclear recoil is the simplest way of conveying high kinetic energies to atoms and ions.

Hot atoms represent very peculiar and reactive species in chemistry. As a result of their initial translational energy, which is much higher than chemical bonding energy, hot atoms disrupt the molecules with which they collide. In the gas phase, however, the hot atom can easily replace an atom from the struck molecule, as in a billiardball type of collision; an alternative reaction is the extraction of an atom from the target molecule, resulting in the formation of a free radical.

The best-understood hot reaction in the gaseous phase is that of tritium, the radioactive isotope of hydrogen. "Hot" tritium is usually formed by bombardment of ^{3}He with thermal neutrons:

$$\text{helium-3} + \text{neutron} \rightarrow \text{tritium} + \text{proton}$$

At the instant of its formation, the tritium atom (represented by the symbol T) has a recoil energy of 200,000 electronvolts. With methane (CH_4), the simplest hydrocarbon, the two typical reactions of hot tritium are the replacement of a hydrogen atom by tritium:

$$CH_4 + T \rightarrow CH_3T + H$$

resulting in methane labeled with tritium, and the extraction of a hydrogen atom with formation of radioactive molecular hydrogen:

$$CH_4 + T \rightarrow HT + {}^\bullet CH_3$$

whereby a residual methyl-free radical is formed.

Reactions of hot carbon atoms have also been extensively investigated both with the long-lived ^{14}C (5730 years) produced in the $^{14}N(n, p)^{14}C$ reaction and the more convenient ^{11}C (20 minutes). The high energy carbon lacks several electrons at its birth and rapidly undergoes chemical reactions. One is the insertion between a carbon-hydrogen bond, for example, with ethane,

$$^{11}C + CH_3CH_3 \rightarrow CH_3CH_2{}^{11}CH^{\bullet}$$

The formed radical rearranges to propane, $CH_3CH_2{}^{11}CH_3$. In this way the target molecule is enlarged by one carbon atom and the resulting free radical is stabilized through further reactions. Cosmic hot carbon atoms probably play an important role in the interstellar synthesis of organic molecules.

In a liquid, mobility is hampered by the surrounding molecules and a hot atom cannot be effectively separated from the trail of reactive fragments it has produced while slowing down. This limitation is described as a "cage effect", and is due to molecules which form a kind of wall around the hot atoms. The reaction of the radioactive atoms with the radical debris is very probable, and a variety of compounds incorporating the radioelement may be produced. As a result of the cage effect, the separation yield of ^{128}I in the Szilard-Chalmers reaction is below 100 percent.

In a solid system the cage effect is still more pronounced because of a denser packing of atoms. The energy of the hot atom is dissipated over a short range in a small volume, playing a kind of havoc commonly designated as "radiation damage".

When a hot atom hits a partner in a crystal lattice, the struck atom may be removed from its position if it acquires an energy higher than about 25 electronvolts. In turn, the displaced atom strikes further atoms in a cascade-like perturbation.

The hottest atoms are the fission fragments, each of which has a kinetic energy of about 100 megaelectronvolts. The associated radiation damage is particularly severe in nuclear fuel, where the range of the hot fission atoms is about 0.01 millimeter. For each fission reaction, between a hundred and a thousand million atoms are displaced in a small volume constituting a "thermal spike". It is considered that the material within this region is molten for a very short time and is then rapidly tempered; after cooling, about 1000 atomic displacements still remain for each fission event. The radiation damage on the microscopic scale produces a mechanical deformation, resulting in brittleness of the fuel material. Chemical and mechanical damage can be repaired to some extent by controlled heating.

Quite often the overall result of thermal annealing appears as an "isotopic exchange" reaction, i.e., a chemical reaction in which the atoms of a given element interchange between two or more chemical forms of the elements. This effect would remain undetected without the aid of radioactive tracers. A typical case in a liquid is the isotopic exchange of iodine atoms between ethyl iodide (C_2H_5I) and iodide ions (I^-) in the liquid phase:

$$C_2H_5\,I + I^{*-} \rightarrow C_2H_5\,I^* + I^-$$

Hot Atoms in Nature

Spontaneous nuclear recoil effects proceed as immutably as the radioactive decays. The action of recoil atoms on terrestrial materials has been continuing ever since the formation of our planet and when we consider the 4500 million years of existence of the Earth, it appears that the cumulative effect should be quite appreciable. However, many of the changes once brought about have since disappeared, owing to various annealing processes such as weathering or tectonic activities at the Earth's surface.

A few witnesses of the accumulated damage due both to the recoil atoms and to radiations are still encountered in minerals containing long-lived natural radioelements or in ones which once incorporated larger amounts of shorter-lived radionuclides that are now extinct.

Color changes are visible effects of radiation damage in ionic crystals. Irradiated rock salt is brownish, the actual tint depending on the radiation dose delivered to the material and on the presence of impurities which behave as color centers. In the laboratory, intense colorations are observed when common salts are irradiated with γ-rays: potassium chloride takes on a deep violet color, sodium chloride becomes yellow, and cesium chloride becomes sky blue. Depending on the sample and temperature the coloration may fade after millions of years or within a fraction of a minute.

A group of heat-resistant and transparent minerals called micas contain radioactive inclusions. The recoil atoms produce circular colored zones around the inclusion, which can be easily observed because of the transparency of the mineral. Several black-brown rings may be distinguished, with radii corresponding to the range of the various α-particles emitted by the radioelement. These rings are known as pleochroic halos because, when exposed to light, they show different colors depending on the orientation of the crystal.

A remarkable effect of radioactive decay in minerals is the transformation of their crystalline structure into an amorphous phase. This phenomenon is most strikingly manifested in zircons and is termed metamictization. It was observed more than a century ago, but its origin was only recognized after the discovery of radioactivity. The β- and γ-radiations produce mainly ionizations which do not affect the crystalline structure. On the other hand, a heavy α-particle towards the end of its trajectory displaces 100 to 1000 lattice atoms in each collision. Even more numerous are the displacements caused by the recoiling daughter nuclide. Generally, the effect increases with the amount of radioelements in the sample.

Atomic displacements are quite common in radioactive materials. For example, in a compound of ^{210}Po (140 days half-life), each atom is displaced on an average of once in a day. The crystallographic order may be further disorganized by changes in the charge and size of the daughter atom with respect to the parent nuclide.

Recoil phenomena also play an important role in geochemistry, and this fact may be of concern in the precision of radioactive clocks and in dating geological events. Although the heavy natural radioelements are initially confined in specific minerals, the daughter nuclide is frequently found spread over large areas. The recoil processes perturb accumulation of atoms in the radioactive decay series and in combination with weathering effects this may strongly affect the equilibrium activities. A good example is provided by the sequence of events relevant to uranium:

$$^{238}U \xrightarrow{\alpha} {}^{234}Th \xrightarrow{\beta} {}^{234}Pa \xrightarrow{\beta} {}^{234}U$$

In a mineral which is well isolated from its surroundings, the two uranium isotopes are in equilibrium and they have the same activity. However, in rocks and ground waters the observed ratio of the activities of ^{234}U to ^{238}U is quite often different from unity and may have values ranging from 0.5 to 120. This represents the largest degree of isotope disequilibrium in nature.

In the isotope fractionation several mechanisms are involved, but they are all engendered by the α-decay of ^{238}U. The ^{234}Th daughter atom has a recoil range of about 50 nanometers in a solid material. It may be ejected from the surface and decay further to ^{234}U in the surrounding system, for example, in groundwater. If the recoil atom is produced deep inside the material, its trajectory ends in a disturbed zone and is stabilized at more labile sites than the parent atoms. ^{234}U, the product of the subsequent decays of ^{234}Th, originates in a perturbed environment and is more easily leached by groundwater than neighboring atoms in rigid regular lattice sites. In oxygen-containing minerals, this nuclide is also likely to be oxidized to the water-soluble hexavalent uranyl cation, while the parent ^{238}U is predominantly in the tetravalent, insoluble form.

The leachability of radioelements from minerals varies widely. Monazite, a mineral containing uranium and thorium, shows exceptional stability to radiation damage even after thousands of millions of years. Such materials are considered as model compounds for the incorporation of radioactive wastes, in particular those containing α-emitting transuranium nuclides.

Applied Hot Atom Chemistry

In the early days of artificial radioactivity, the Szilard-Chalmers effect was used for the concentration of radionuclides and their production with high specific activity and a minimum amount of carrier material. The procedure is still used in a limited number of cases, such as for the preparation of ^{32}P, ^{51}Cr, and ^{64}Cu.

The most important present application of hot atom reactions is the production of radiopharmaceuticals, which are compounds containing short-lived radionuclides used in medical research and clinical applications (Chapter 8). For *in vivo* studies, these materials have many advantages: a low dose is delivered to the patient, the amount of carrier is in the nanogram range, the biological equilibrium remains undisturbed, and even toxic compounds can be injected. Moreover, the use of short-lived isotopes avoids the waste problem.

The most widely used short-lived isotopes in medicine are ^{11}C (half-life 20.3 minutes), ^{13}N (10 minutes), ^{15}O (2 minutes), and ^{18}F (110 minutes). These cyclotron-produced nuclides are all positron emitters and are thus convenient for the PET techniques (positron emission tomography, Chapter 8). The fast synthesis of molecules containing the radioelement can be achieved by hot reactions between the nascent radionuclide and a suitable target molecule, leading to a convenient precursor for more elaborate syntheses.

^{11}C is obtained by proton bombardment of nitrogen gas mixed with appropriate additives. Depending on the latter, ^{11}C is obtained as ^{11}CO, ^{11}CN, $^{11}CH_4$, $H^{11}CN$, or $^{11}CO_2$, all of which are reactive species for further syntheses. Relatively complex labeled molecules such as glucose, fatty acids, amines, and steroids can be made in a few minutes.

The bombardment of nitrogen mixed with O_2 is used to produce labeled oxygen, $^{15}O_2$, and nitrogen oxides. ^{13}N atoms obtained in the reaction of protons with oxygen at high pressure are incorporated in $^{13}N_2$ and $^{13}NO_2$. In the presence of water, part of the ^{13}N is found as $^{13}NH_3$, $^{13}NO_2^-$, and $^{13}NO_3^-$.

Hot atoms are the very essence of nuclear energy. In fission reactors they determine the chemical behavior of the fission products and their release from the nuclear fuel rods. Tritium, which is the fuel of fusion reactors, is formed by neutron irradiation of lithium. The recovery of tritium from a blanket made of lithium salts, and recognition of the nature of the radiation damage caused to the construction materials by the recoil atoms are factors of primary interest in the successful operation of a fusion power plant.

Hot Atoms in Space

In space, hot ions or atoms are ubiquitous and abundant. An intense source of accelerated species is the Sun, with its solar winds and flares. Solar wind particles have velocities of 400 kilometers per second corresponding to about 10,000 electron-volts for carbon ions. A further contribution to hot species in space arises from acceleration processes in the radiation belts of planets and in cometary comae. Cosmic radiation and decay of radionuclides provide a continuous source of hot species. Hot reactions are also produced by collisions of interstellar gas and dust clouds, shock waves, and expansion of remnants of supernovas.

The targets for hot reactions in space comprise mainly gaseous or solid systems, such as interstellar gas and clouds of dust. Interstellar grains consist typically of rocky cores with diameters of the order of 0.1 micrometer, with layers of organic refractory compounds and frozen volatile compounds such as water, ammonia, and methane (Chapter 6).

The nature of reactions involving hot species in space has been inferred from computer simulation and from laboratory experiments.

CHEMICAL EFFECTS OF RADIOACTIVE DECAY

Beta Decay

When a radioactive nucleus disintegrates, its chemical identity changes. In the case of β^- radioactivity, the atomic number increases by one unit and the atom formed in the decay is shifted to the next higher, adjacent position in the periodic classification of the elements. Thus tritium decays to helium, ^{14}C to nitrogen, and ^{32}P to sulfur. Nucleogenic atoms, i.e., atoms formed in a nuclear decay, not only have altered chemical properties, but are also born with characteristic excitation and kinetic energies.

β-decay, like any radioactive disintegration, is a random process, but when it occurs it proceeds rapidly. The nucleus suddenly acquires an additional proton with the result that its positive charge increases by one unit. Consequently, the attractive force on the electrons increases also, and the latter are slightly shifted towards the center of the atom. In some instances, the electronic shells do not have sufficient time to accommodate themselves to the increased nuclear charge. In this case some energy becomes available for excitation. The electronic shells are "shaken" and it may happen that several electrons are detached from the atom, which then becomes an ion with several units of positive charge.

Because of the small mass of the electron, the recoil energy following β-emission is weak, being often of the order of, or less than, the chemical bond energies.

When the radioactive atom is a constituent of a molecule, the chemical change induced by the decay is the most important consequence. Depending on the extent of change in chemical properties, the daughter atom may possess a different valence state and radius and be incompatible with the initial molecular structure. For example, the decay of tritium invariably breaks the molecule, because the daughter atom, helium, is a noble gas which cannot form chemical bonds. In the simple structure of the labeled molecule methane CH_3T, a tritium atom (T) replaces a hydrogen atom H. When tritium decays to helium, the molecule becomes an ion CH_3He^+ which breaks up immediately into $CH_3^{\bullet}$ and He. In many cases, the radical $CH_3^{\bullet}$ is further fragmented by virtue of the excitation energy which has become available in the decay.

.   However, the newborn atom may remain in the molecule when its chemical properties do not differ significantly from those of the parent atom. In this way, novel chemical species originate from radioactive decay. For example, in the disintegration of ^{14}C to ^{14}N, the latter species has a fair chance of remaining in the original molecule, simply substituting the parent carbon atom. The decay of ^{14}C incorporated in methane ($^{14}CH_4$) leads to the stable ammonium ion ($^{14}NH_4$).

More elaborate decay-induced syntheses have been demonstrated in the laboratory. An important component of many biomolecules is the carboxyl group, –COOH, which is a characteristic part of organic acids. Labeled ^{14}–COOH, after decay of the radiocarbon, rearranges to the amino group NH_2, another chemical function group in numerous biological compounds. These processes may lead to molecules containing both the COOH and NH_2 groups; such compounds are known as amino acids, the basic constituents of life.

Similar processes are probably operative in nature. Radiocarbon is an omnipresent nuclide which is continuously generated in nuclear reactions produced by cosmic rays. When it decays, ^{14}C is transmuted to nitrogen and many types of biologically important molecules can be synthesized in this way. Knowledge of the abiotic formation of these molecules on and outside the primitive Earth is of great interest in understanding processes of prebiotic evolution.

In a similar manner, the incorporation of phosphorus atoms into biomolecules can be attributed to the decay of ^{31}Si. The latter is produced by the reaction of cosmic neutrons with silicon, which is the most important element in the soil. Laboratory experiments have proven the abiological synthesis of adenosine monophosphates using ^{31}P atoms produced by the decay of ^{31}Si. On the same lines, the conversion of alanine to cysteine, a sulfur-containing amino acid, has been accomplished on the basis of ^{32}S resulting from the decay of radioactive ^{32}P.

Chemical effects induced by decay also occur in the human body because every biomolecule (and, by extrapolation, every living cell) contains a small number of radioactive atoms, in particular ^{14}C and tritium of cosmic origin. Other radionuclides, such as ^{32}P, are transported in chemically bonded form through the food chain to man. It is important to have an understanding of the nature and chemical and biological behavior of the new molecules which are formed subsequent to the decay of radioisotopes. Effects of transmutation add to those of the radiolytic decomposition of the medium that is directly induced by radiations. It is often difficult to assess the respective role of both. Extensive research has been conducted on the decay effects of tritium and radiophosphorus in the DNA molecule, which is a basic constituent of living organisms. The skeleton of DNA consists of two helicoidal strands. Among other, less violent changes, transmutation can induce a break in one or both strands. A single-strand break can be efficiently repaired by the living cell, whereas double-strand breaks generally cannot and may have lethal consequences. Faulty repair induces chromosomal abberations, which eventually cause mutations.

At our present state of knowledge it appears that the death of a cell results from the break of the two strands owing to radiolysis induced by the β-particles of the incorporated nuclide. Mutagenesis, however, is a genuine local transmutation effect due to a rearrangement of the parent molecule and an ensuing change in the genetic code. The latter effect is strongly dependent on the position of the radioactive atom in the DNA unit.

Electron Capture and Internal Conversion

In some instances, the chemical effects of a radioactive decay may be particularly violent, for example, when the process leads to the ejection of an electron from a shell close to the nucleus. This always happens in electron capture decay, in which an electron is captured by the nucleus itself; such ejection can also occur in an isomeric transition when the available decay energy is used to expel an inner electron. Both events leave an electron vacancy deep inside the atomic core, but this is immediately filled by an electron from the next higher shell. This transfer releases enough energy to eject a further electron from the atom, thus resulting in an additional electron vacancy in the shells close to the nucleus. The process thus triggered by the disintegration follows its course, and more and more electrons are ejected. Eventually the avalanche-like ionization attains the outermost or valence shell, which is responsible for chemical bonding in a molecule. It has been found that an atom can suffer a loss of up to 20 electrons in this way.

What happens to the molecule when one of its atoms acquires a high buildup of positive charge in this manner? It is considered that, in the gas phase at least, this atom acts as a sink for electrons from the neighboring atoms. The resulting electron migration causes positive charges to build up at the extremities of the molecules. From the effect of Coulomb repulsion the molecule is likely to split apart, whereby the fragments acquire a kinetic energy of up to several electronvolts. Measurements have confirmed that the molecule is totally ruptured following electron capture decay.

In a condensed medium, in which the molecules are closely packed, it is likely that electrons are transferred to the positive center from neighboring molecules. In any event, however, the electrons released in the process of multiple ionization dissipate their energy within a distance of about 10 to 100 nanometers and effectively damage molecules which are close to the site where the radionuclide has decayed.

Thus, transmutation affects not only the molecules in which the radioactive atom was initially bound, but also its environment. The drastic biological effects of the electron-capture decay of ^{125}I have been observed in biomolecules and in living cells. When the nuclide is incorporated into the vital DNA, the strands are broken over a distance of several nanometers from the decay site. Accordingly, ^{125}I is one of the most deadly cell killers; its action at the molecular level might possibly be useful for microsurgery and its therapeutic use for the treatment of malignant neoplasms has been suggested.

Dependence of Nuclear Processes on the Chemical Environment

In chemical reactions, electrons are transferred from one atom of the molecule to another and the nucleus does not intervene in this exchange. Similarly, when a nucleus is transmuted, the electrons generally do not participate in the process. This rule has some exceptions, namely, when the atomic electrons are involved in the radioactive decay, such as in the case of electron capture. Chemical binding affects mainly the outermost shells. Thus, to the extent that outer electrons may be captured by the nucleus, the probability of the decay and, hence, the half-life of the radionuclide will be influenced. The effect is more pronounced in light atoms, but, in any event, the change in the decay rate as a function of the type of chemical bonds does not exceed a few parts in 10,000 and is therefore very difficult to determine. One of the few nuclides for which such a change has been detected is ^{7}Be, which has a half-life of 53 days. The difference between the decay rate of metallic ^{7}Be and of the same isotope in various compounds is about 0.1 percent.

Excited nuclei return to the ground state by emission of an electromagnetic radiation, the γ-rays. Although the origin of the latter lies in the nucleus, it has been found that the exact energy of the radiation depends on the chemical state of the corresponding atom. The variation is extremely small, but it can be determined with a degree of precision comparable to that of measuring the distance between the Earth and the Moon (400,000 kilometers) with a precision of a fraction of a millimeter!

Further Reading

Adloff, J. P., Gaspar, P. P., Imamora, M., Maddock, A. G., Matsuura, T., Sano, H., and Yoshihara, K., Eds., *Handbook of Hot Atom Chemistry*, Kodansha, Tokyo/Verlag Chemie, Weinheim, 1992. [The most recent encyclopedic review in the field, also written by an international team of specialists.]

Friedlander, G., Kennedy, J.W., Macias, E.S., and Miller, J.M., *Nuclear and Radiochemistry*, 3rd ed., John Wiley & Sons, New York, 1981. [Covers all chemical aspects of nuclear processes.]

Stoecklin, G., *Chemie heisser Atome, Chemische Reaktionen als Folge von Kernprozessen*, Verlag Chemie, Weinheim, 1969. [An excellent short introduction into the topic. Highly recommended for its pedagogic value. Translated in French (Masson, Paris).]

Tominaga, T., and Tachikawa, E., *Modern Hot-Atom Chemistry and Its Application*, Springer-Verlag, Berlin, 1981. [Written for the nonspecialists, with more emphasis on application of hot atom chemistry in related fields.]

Matsuura, T., Ed., *Hot Atom Chemistry*, Kodansha, Tokyo/Elsevier, Amsterdam, 1984. [An up-to-date advanced review written by an international team of specialists.]

Radiation, radioactivity and nuclear energy are intrinsic properties of matter. We have already seen how these aspects are investigated in the laboratory (Chapters 2 to 4). This chapter shows the facilities that the universe itself provides for such studies. Basic information on the present state of the universe is given, and an overview of data is provided with respect to an event which began some 15,000 million years ago with a "Big Bang".

Modern concepts of the nature, origin and history of the universe are based essentially on two findings: first, the galaxies are receding in a manner that suggests flight from a central point with steadily increasing velocities; second the universe literally bathes in a sea of microwave radiation. Recession of the galaxies implies an explosive beginning and the temperature of the fossil radiation (2.7 degrees kelvin), after thousands of millions of years of cooling, shows that a superhot fireball must have exploded during the "Big Bang".

A detailed theory of the course of events in the early universe is widely accepted as the "standard model". It offers valuable insight into both the evolution of the universe and the nature of matter, especially with regard to its elementary constituents and basic forces.

Nucleosynthesis in the early universe produced the nuclei of deuterium and helium. Once the stars were formed, violent nuclear processes in the stellar cores initiated the syntheses of nuclei heavier than hydrogen and helium, and these processes still operate. The explosions of supernovas are the sites of syntheses of nuclides of the heaviest chemical elements.

Supernova explosions also contribute to the formation of cosmic rays, the radiation that fills interstellar space and bathes all celestial bodies. It is a mixture of charged particles and electromagnetic radiations of various energies. Protons are the predominant constituents, and the particle composition reflects that of the elemental abundance in the universe. The energies and fluxes of cosmic rays vary enormously.

Studies of antimatter in the laboratory are still at a primary stage. Potential military aspects already seem frightening.

Invisible dark matter dominates the material content of the universe. Its true nature is still unknown and it represents a challenge for cosmologists and particle physicists.

5

Nucleosynthesis, Cosmic Radiation, and the Universe

A LOOK BEYOND THE EARTH

The distances are enormous — this is the first thing to realize when looking for celestial bodies: about 150 million kilometers to the Sun, 40 times as far to reach Pluto, the most remote member of the solar system, and about a 100,000 times farther to reach the nearest star.

If we look farther out into cosmic space, the distances become even greater and more difficult to express in conventional units. Astronomers use a unit called the light-year, which is the distance that a light ray, or any electromagnetic radiation, travels in 1 year; it is 9.46 million million kilometers (9.46×10^{12} kilometers). The most distant galaxies are more than 10,000 million light-years away. Some astronomical objects called quasars may lie at distances of 16,000 million light-years.

It is noteworthy that the light which corresponds to such distances left these celestial bodies so many thousand million years ago, and thus their observation reveals not what is going on at present, but what has happened in the universe in the past, backwards towards its beginning.

Table 12

SOME PROPERTIES OF ELECTROMAGNETIC RADIATION OF COSMIC ORIGIN

Radiation	Energy range in electron-volts[a]	Wavelength in meters[a]	Source
γ-rays	from 10^5	below 10^{-11}	Pulsars
X-rays	10^3—10^5	10^{-11}—10^{-9}	Black holes
Ultra-violet	6—10^3	10^{-9}—2×10^{-7}	Hot stars
Visible	1—6	2×10^{-7}—10^{-6}	Cold stars
Infrared	0.01—1	10^{-6}—10^{-4}	Planets, interstellar dust
Microwave	10^{-5}—10^{-2}	10^{-4}—10^{-1}	Interstellar molecules, quasars
Radiowave	below 10^{-5}	from 0.1	Galaxies

[a] The distinction between the various kinds of radiation is not sharp

The human eye recognizes only a tiny fraction of the real universe, namely, that which is revealed by visible light. This is a very narrow window in the electromagnetic spectrum (Table 12). Much more abundant are the signals from the "invisible" universe, which extend from radiowaves to gamma-rays, including the infrared and ultraviolet radiation and X-rays. Most of these radiations are absorbed by the atmosphere, but those which reach the surface of the Earth, or can be captured by instruments carried by rockets and satellites, make the joy of astronomers.

Our geocentric vision of the universe, from the terrestrial sphere with a radius of 6357 kilometers, encompasses the Moon at a distance of 1 light-second, the sun at 8 light-minutes, Pluto at 5.3 light-hours, and a multitude of stars of which the closest is a little more than 4 light-years away and Sirius at 8 light-years.

The Solar System is about 11 light-hours across. It is imbedded in a conglomeration of stars, the giant disk-shaped galaxy called the Milky Way. We see only a part of this as a band of diffuse light that stretches across the heavens on a clear and moonless night, but a detailed picture can be obtained from various studies of intra- and extragalactic space. The center of the galaxy cannot be seen with optical telescopes, but present-day radiotelescopes penetrate to a distance of 30,000 light-years from the Earth, corresponding to the galaxy's periphery.

The galactic disk measures 100,000 light-years from edge to edge. It is composed of interstellar gas and dust, 100,000 million stars, and, most likely, 1000 million planets and their satellites. Its central bulge, within a diameter of about 10,000 light-years, consists of densely packed older stars. A thin disk around the galactic nucleus contains stars of all ages, including those just born and others in the process of expiring.

The total mass of our galaxy, including the disk and the bulge, is estimated as well under 200,000 million solar masses. The disk and bulge are surrounded by a "halo" of matter on each side of the galaxy's central plane, perhaps adding a further 100,000 million solar masses to the total mass of the galaxy. All these celestial objects are arranged in a pinwheel pattern of spiral arms that revolve around the galactic center. Speeding along at about 200 kilometers per second, the Sun takes 230 million years to circle the galaxy once, during the so-called galactic year.

In addition to stars, white spots with a cloudlike appearance are also visible in certain parts of the sky. Some of these consist of clouds of gas and dust within our galaxy, where important chemical reactions take place which are induced by ionizing radiation. However, some of these bodies are galaxies in their own right, such as the Andromeda Nebula, which is the closest of its kind to the Milky Way and lies at a distance of 2 million light-years. On the whole, the universe probably includes 10,000 million galaxies which, on the average, are separated from each other by at least 1 million light-years.

Certain celestial objects resembling stars can be observed at distances greater than that of the farthest galaxy yet identified. Each of these emits more energy than our own galaxy. Their true nature is virtually unknown, and because of their quasi-stellar appearance they are called "quasars".

Another group of curious astronomical objects includes the "black holes". They cannot be directly observed but most astronomers are convinced of their existence. They are thought to be the remnants of dead massive stars, comprising a form of collapsed matter in which pressure, density, and gravitation are so high that even the most energetic particles and electromagnetic rays cannot escape. They thus remain invisible.

THE EXPANDING UNIVERSE AND FOSSIL RADIATION

Two important findings are decisive in our present approach to the nature, origin and evolution of the universe: first, the galaxies are rapidly receding, and second, cosmic space bathes in a sea of microwave radiation. These facts have led to a now widely accepted "standard model" which is useful not only for cosmology, but also for an understanding of the nature of matter, its elementary constituents, and the basic forces which operate throughout the range from celestial to subatomic levels.

Observations

The motion of stars is inferred from spectral analysis of the light that they emit; such measurements are based on a phenomenon known in physics as the Doppler effect. The latter states that the frequency of a wave increases when the wave emitter is displaced towards the observer, and decreases when it moves in the opposite direction. This effect holds not only for light but also for all other types of waves, including those of sound: it is a common experience that a car's horn sounds shrill when the car is approaching and is low pitched when it is moving away.

Because of the Doppler effect, the frequency of light emitted by atoms in receding stars is lower than that recorded for the same atoms in the laboratory. In other words, the frequency is shifted toward a longer wavelength. Since this corresponds to a displacement toward the red part of the visible spectrum, it is said that the light emitted by a star which is moving away from us exhibits a "red shift". The farther away the celestial object is, the greater the extent of red shift.

The celestial bodies are rushing apart as if they were moving from a central point with steadily increasing velocities. The farthest known object in the universe, a quasar, has a red shift corresponding to a recessive velocity of 92 percent of that of light.

The increase in velocity of recession is constant for a given distance. Its evaluation may vary considerably because of experimental difficulties in determining the red shifts of remote galaxies. At present, the best value is considered to be 17 kilometers per second per million light-years.

In 1965, a microwave radiation background was discovered in the universe. This observation was made with a radiotelescope, or rather radioantenna, which recorded photons with a wavelength of 7.35 centimeters. This effect is usually referred to as "2.7 degrees kelvin thermal radiation". In the search for 7.35-centimeter radiation, the radiotelescope was aimed at chosen positions in space. The recorded radiosignals showed that the radiation arrives with equal intensity from all directions in space, i.e., it is isotropic.

After more than 2 decades of observations, astronomers have still not detected any variation in the intensity of radiation across the sky, except that due to the relative motion of the Earth. Experiments in space are carried out to detect minute fluctuations in the microwave background. The scientists hope that the findings will help to explain how the early universe evolved.

The Beginning

Since the galaxies are moving away from us, they must have been closer together in the past. One can calculate the time required for any pair of galaxies to reach their present degree of separation by taking into account the constant increase of the velocity of recession, 17 kilometers per second per 1 million light-years. Since 1 million light-years = $10^6 \times 9.5 \times 10^{12}$ kilometers, the time needed for two galaxies to separate from a common point is 9.5×10^{18} kilometers divided by the velocity 17 kilometers per second:

$$\frac{9.5 \times 10^{18} \text{ km}}{17 \text{ km/second}} = 5.59 \times 10^{17} \text{ seconds} = 17,740 \text{ million years}$$

This value for the age of the universe is called the Hubble time after Adwin Powel Hubble (1889—1953), the astronomer who first observed recession of the galaxies and who performed the first red shift measurements. Because of uncertainty in the value of the constant of recessive velocity increase, the Hubble-time of the universe is usually stated as (15,000 ± 3000) million years.

The fact that the birth of the universe is established with some confidence leads inevitably to the question: if there was a beginning, then what was before? Nothing? Nothing can engender only nothing. The question remains open and the interested person might be referred to Saint Augustine who, when asked what the Lord was doing before He created Heaven and Earth, apparently replied: "He was creating hell for people who ask such questions." To elude the problem of the beginning, some scientists have suggested that the universe is eternal, but many of their assumptions are not consistent with observations.

There is, nevertheless, a logically acceptable approach. Time is counted from the beginning, the "absolute zero time", and time itself has no meaning before that moment. A similar approach is generally accepted in the case of the absolute zero temperature: it is impossible to cool anything below -273.16 degrees celsius because the concept of temperature for values lower than the absolute zero has no physical meaning. In the same way, we may have to get used to the idea of an absolute zero time: a moment in the distant past beyond which it is in principle impossible to trace any relation from cause to effect. The question, of course, is a matter of conjecture.

Curiously enough, the notion of a singular primordial event from which the universe was born satisfies not only a large part of the scientific community, but also the poet and philosopher

(Goethe's Faust: "Am Anfang war die Tat") and the believer (Genesis I.l: "Fiat lux, et lux fiat").

ENERGY AND MATTER IN THE UNIVERSE

The concept of a singular primordial event enables us to sketch the broad outlines of the evolution of the universe starting from an undifferentiated mixture of energy and radiation and continuing up to the present day. The process starts with the explosion (a "Big Bang") of an infinitely small volume. In the expanding and cooling fireball, and after a small fraction of a second, the unified force separates into four fundamental natural forces. Within the first second other important events occur: the elementary particles separate from one hypothetical, fundamental constituent of matter; the annihilation reactions of matter and antimatter cease, leaving an excess of matter; the nuclei of hydrogen appear. The formation of nuclei of deuterium and helium is complete within the first few minutes. The first atoms, those of hydrogen and helium, appear after 300,000 years. This marks the end of the stage which started with the Big Bang, but the evolutionary process continues. In the expanding and cooling universe the matter accretes, and the first galaxies and the stars appear after 1000 million years. In the stellar cores the nucleosyntheses take place and heavier atomic nuclides form.

The model considers the evolution of energy and matter on an extraordinarily large scale of time (10^{-43} to 10^{20} seconds), temperature (10^{32} to 2.7 degrees kelvin), and density (10^{50} to 10^{-30} grams per cubic centimeter). There are many arguments in its favor. The formation of galaxies is not observed at present, whereas that of stars is, as well as the recession of both. The omnipresence of the cosmic ray background and the constancy of the deuterium to helium ratio also strongly support the general picture. Various theoretical assumptions with respect to fundamental forces and elementary constituents of matter are being confirmed by experiments in laboratories.

The broad outlines are widely accepted by scientists, and the model is used as a powerful tool in cosmology and high-energy physics. However, not all theoreticians are in agreement on specific details. This is particularly true for the important events that occur on the subsecond time scale. Depending on the parameters used in modeling, some findings may vary by several orders of magnitude, such as the time dependence of temperature and density. The present description is, therefore, only one of likelihood and it will be subject to change as research progresses.

Forces and Elementary Particles

The universe starts as a zero volume of infinite density. This state is called a singularity. A unified theory of elementary particles and forces assumes that it consisted of one elementary constituent of matter (unifying the elementary particles) and one single force (unifying the four forces now prevailing in nature).

The model does not account for the origin of the singularity. Moreover, the process begins within a small fraction of a second after the Big Bang. After 10^{-43} second the gravity force separates and becomes independent from the other three (strong, weak, and electromagnetic). The strong force emerges as an independent entity after 10^{-35} second. It causes a rapid expansion of the universe (between 10^{-35} and 10^{-32} second) without involving appreciable cooling; this is the so-called inflationary period. Adding such a period to the theory helps to solve some problems of the classical Big Bang cosmology. It also offers new possibilities such as the existence of cosmic strings. These are one-dimensional relics of the early universe with a huge mass per unit length; 1000 million tons per centimeter.

Within 10^{-35} second, another important event occurs. Most of the matter and antimatter undergo mutual annihilation, but the rapid expansion somehow leaves a small excess of matter. Its evolution leads, some 10^9 years later, to the galaxies, stars, and all the matter around us.

The last unified force (electroweak) divides into electromagnetic and weak interaction at $t = 10^{-10}$ second and at a temperature $T = 10^{15}$ degrees kelvin.

About 1 microsecond after the Big Bang, the temperature decreases sufficiently ($T \cong 10^{13}$ degrees kelvin) for the annihilation processes to cease. The quarks (Chapter 2, Table 5) become locked into mesons (Chapter 2, Table 6), and baryons (Chapter 2, Table 7). The binding of quarks is ensured by the exchange of gluons, the force-carrying particles (Chapter 2, Table 3). In reality, these physical processes are not so simple. The confinement of quarks releases enormous amounts of energy, which in turn can again cause separation of the particles, and thus the synthesis proceeds back and forth until the quarkian storm has subsided. The nuclei of hydrogen, the protons, are formed before the first microsecond has elapsed.

By the time the universe is 1 second old, its temperature is about 10^{10} degrees kelvin, and the light elementary constituents of matter, the leptons (Chapter 2, Table 4), emerge.

Nuclei of Deuterium and Helium

The primordial nucleosynthesis starts after about 100 seconds, when the universe is sufficiently cool ($\cong 10^9$ degrees kelvin) to allow a proton and a neutron to merge into a new nucleus: the deuteron, a heavy isotope of hydrogen. The deuterons accumulate since the energy of most of the photons at a temperature of 1000 million degrees kelvin is not high enough to cause splitting back into nucleons. The deuteron captures one or two more nucleons to build the nuclei with atomic masses 3 and 4. Alternatively, two deuterons amalgamate to form a four-nucleon species: the nucleus of ^{4}He.

The complete conversion of hydrogen to helium in the primordial nucleosynthesis is prevented by the lack of neutrons which, when free, disappear by radioactive decay to form protons. Another obstacle is the inability of two protons to form ^{2}He: the electromagnetic repulsion between two protons is superior (by only 2 percent) to their attraction by the strong force.

Formally, nucleosynthesis should continue by adding a neutron to ^{4}He, or merging two ^{4}He nuclei into a species with eight nucleons (^{8}Be). However, as a consequence of the extraordinary stability of ^{4}He, the products of these reactions decompose rapidly: 10^{-21} second suffices for ^{5}He to release the neutron, and 10^{-16} second for ^{8}Be to separate back into two ^{4}He nuclei. Formation to a minor extent of ^{7}Li is due to a combination of three protons and four neutrons.

After 250 seconds have elapsed, the mass of the universe consists of about 75 percent hydrogen (protons and deuterons), nearly 25 percent helium (^{3}He and ^{4}He), and a trace of ^{7}Li. The overwhelmingly predominent nuclides are the proton (1000 protons for 1 deuteron) and ^{4}He (10,000 ^{4}He for 1 ^{3}He).

Predictions of the abundances of light nuclei due to the primordial nucleosynthesis provide a convincing test of the standard model.

The First Atoms and Molecules

The products of nucleosynthesis during the first few minutes are the nuclei of light atoms, bathing in a sea of photons, electrons, and neutrinos for several thousand years. The ambient temperature is too low for nucleosynthesis to continue and too high for a synthesis of atoms and molecules to start.

The situation changes appreciably after about 300,000 years, when $T \cong 3000$ degrees kelvin and the Coulomb attraction becomes sufficient for electrons to be bound to the nuclei of hydrogen and helium. This corresponds to the birth of the oldest atoms in the universe, those of hydrogen and helium.

At a further stage of expansion and cooling follows the first chemical synthesis: the binding of two hydrogen atoms into a molecule of hydrogen, H_2.

Once the atoms and molecules were formed they began to cluster, and eventually after about 1000 million years participated in the formation of galaxies and stars. The main raw material for the formation of stars and the continuation of nucleosynthesis in the stellar core is hydrogen. Fusion "burns" the nuclei and, as the nuclear fuel becomes depleted, the stars age and eventually die. The physical characteristics of a star during its lifetime, as well as its death, are the consequences of nuclear processes in the star's core.

Eras Dominated by Radiation and Matter

At present, the 2.7-degree kelvin background radiation consists of about 550 million photons per cubic meter. This is a high density when compared to that of nucleons, which is estimated at 0.03 to 6 per cubic meter.

The number of photons exceeds by 100 million to 20,000 million times that of the particles, depending on the actual value for particle abundance. And yet we live in an era dominated by matter. The reason is that the mass at rest of a nucleon is about 939 megaelectronvolts, while that of the 2.7-degree kelvin photon is only 0.00026 electronvolt. A simple calculation gives for the mass content due to photons 0.143 megaelectronvolt per cubic meter, and from 5.63 to 0.03 gigaelectronvolt per cubic meter due to particles (expressed as energy equivalent).

In the first 300,000 years after the big bang, the temperature of the universe was much higher, as was also the energy of photons. The latter lies in the megaelectronvolt-electronvolt range at temperatures between 10^{11} and 10^4 degrees kelvin. This was the radiation-dominated era, as most of the energy in the universe was then in this form.

The photons became free to travel, and the universe became "visible" after the end of the radiation era (Figure 21).

NUCLEOSYNTHESES IN STELLAR CORES

Stars have a finite lifetime because their supply of energy is limited. The energy is released by a series of nuclear reactions which depend on the temperature in the core. The temperature itself is fixed by the mass and age of the star.

Nuclear reactions not only provide energy, but as the star evolves (particularly in its final stages) they lead to the syntheses of heavier chemical elements from the lighter ones. We shall consider only a few of the most important pathways of nucleosyntheses which have occurred in the oldest stars (population III) or are still taking place (populations II and I).

Helium

Most stars produce energy by transformation of hydrogen into helium and are called hydrogen-burning stars. The temperature

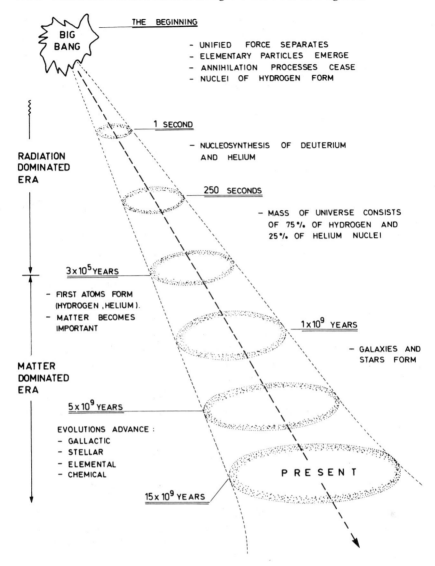

FIGURE 21. Some stages in the evolution of the universe.

in the core must be at least 10^7 degrees kelvin to "ignite" the reaction between the protons. The shining star draws its energy from gravitational contraction until the temperature threshold for nuclear reaction is reached. This change from gravitational to nuclear energy is an important step in stellar life. In the case of our Sun this happened about 10 million years after its birth.

The net balance of the first step in stellar nucleosynthesis is the merging of four protons into a helium nucleus. However, this cannot occur in one step. Even at enormous pressure which prevails inside the Sun's core, the probability that four protons collide simultaneously is insignificant. The formation of a helium nucleus thus has to proceed stepwise. First, two protons must combine to give a deuterium nucleus (with release of a positron). The next step is reaction of deuterium with another proton, producing ^{3}He. The combination of two ^{3}He nuclei gives one stable nucleus of ^{4}He and two protons,

$$^3\text{He} + {}^3\text{He} \rightarrow {}^4\text{He} + 2\ {}^1\text{H}$$

in addition to a considerable amount of energy. This energy (26.7 megaelectronvolts for each nucleus of helium formed) contributes to the maintenance of a high temperature in the Sun's core.

The condensation reaction in which two nuclei merge into a single one is called nuclear fusion or a thermonuclear reaction. The Sun is a fusion reactor which indeed has an astronomical size, and the wall confining the hot plasma consists of external hot layers of the star's surface. The confinement of such a plasma in a volume of a few cubic meters under terrestrial conditions, in an electric power plant based on nuclear fusion, is not an easy task, but encouraging results have already been achieved (Chapter 10).

The formation of each helium nucleus in the Sun's core is accompanied by the emission of two neutrinos, which readily escape into outer space. It is estimated that thousands of millions solar neutrinos per second reach every square centimeter of the Earth's surface.

Carbon

Because of its stability, the helium nucleus is a serious impediment in the nucleosynthesis of heavier elements. However, once a sufficient amount of helium has accumulated, conditions in the core change drastically. The gravitational contraction becomes so important that the density increases up to 10^4 grams per cubic centimeter and the temperature rises to 10^8 degrees kelvin. The Coulomb repulsion between the helium nuclei is overcompensated to the extent that three (not two!) heliums merge into a single nucleus of six protons and six neutrons; this gives rise to the nucleus of ^{12}C.

As before, nucleosynthesis proceeds stepwise. Two ^{4}He give one ^{8}Be, which, although decaying within 10^{-16} second, tends to accumulate. A steady-state concentration is established in which the number of nuclei formed equals that of those which decay; it is sufficient to support the addition of a third ^{4}He to produce ^{12}C.

The helium burning stage is assumed to be the principal energy source in a class of stars called red giants. Nevertheless, in their outer layers the hydrogen burning process may occur simultaneously.

From Oxygen to Iron

The helium burning stage is a relatively short period on the cosmic time scale, ranging from 20 million to 100 million years. As the concentration of ^{12}C builds up, the synthesis of heavier nuclei up to ^{20}Ne takes place by successive capture of ^{4}He. ^{12}C produces ^{16}O which, in turn, gives birth to ^{20}Ne.

After carbon and oxygen nuclei have accumulated to a sufficient extent, renewed gravitational contraction creates the necessary conditions for the fusion of carbon nuclei. The temperature is $(6 \text{ to } 7) \times 10^8$ degrees kelvin and the density 5×10^4 grams per cubic centimeter. Such conditions prevail in stars whose mass is three or more times greater than that of the Sun. The fusion of two ^{12}C nuclei leads to ^{24}Mg, ^{20}Ne, and to several other nuclei of magnesium and sodium.

At a still higher temperature the fusion of two oxygen nuclei, or of an oxygen and a carbon nucleus, leads to a large variety of isotopes of magnesium, silicon, and sulfur. At the end of the ^{12}C and ^{16}O burning stages, the most abundant nuclei are ^{28}Si and ^{32}S.

The fusion of ^{28}Si and the syntheses of heavier nuclei up to mass number 56 occur at temperatures above 10^9 degrees kelvin. The nuclei formed are in the region of iron and nickel in the periodic chart, and belong to the most stable existing nuclei. Their accumulation leads to a new scenario of nucleosynthesis in which photonuclear reactions become involved. Increasingly energetic photons are released in the core as the temperature rises, and these destroy the stable nuclei. Photodisintegration generates neutrons which promote further nucleosynthesis.

Beyond Iron: Reactions with Neutrons

Neutron-induced reactions can occur in the stars of second or later generations, i.e., in those born in the interstellar medium containing the debris of dead stars which have already evolved through advanced stages and contain the chemical elements up to iron.

As a neutral particle, the neutron is insensitive to Coulomb repulsion and penetrates more or less easily into any nucleus.

Nonetheless, the addition of neutrons to a target nucleus cannot be continued indefinitely. After a surplus of one to four neutrons, the equilibrium between neutrons and protons which maintains nuclear stability is disturbed. The nucleus gets rid of the surplus neutron by converting it into a proton and emitting a beta-ray. This is the phenomenon of β-radioactivity which is of paramount importance in nucleosynthesis; from an element with atomic number Z, a heavier element with number Z + 1 is formed, the mass number being increased by the number of captured neutrons. For example, when three neutrons are added to the nucleus of ^{65}Cu (29 protons), the nucleus formed is ^{68}Cu which, after β-decay, becomes ^{68}Zn (30 protons).

This process advances stepwise until the chemical element bismuth is reached, containing 81 protons and 128 neutrons. The long pathway from iron to bismuth includes two radioactive elements: technetium (Z = 43), whose longest-lived isotope has a half-life of 2.6 million years, and promethium (Z = 61) with 18 years half-life. These half-lives are short on the cosmological time scale and preclude the survival of these two elements on our present-day Earth. It is a remarkable fact that spectral analysis of light emitted by some stars has revealed the presence of both these elements. Moreover, they are also found in mixtures of fission products in uranium reactor fuels.

Depending on available targets and their "avidity" for neutrons, reactions in the star cores require a considerable amount of time in comparison with terrestrial standards. Despite the fact that the fluxes are rather impressive, amounting to 10^{16} neutrons per second and per square centimeter, the capture of one neutron requires a time ranging from 100 to 100,000 years. On the whole, the buildup can last for 10 million years or more. This progressive and slow addition of neutrons is called the slow process (s-process).

Nucleosynthesis cannot occur if the nucleus decays before it can capture a neutron. To overcome this obstacle, neutrons must be added on a time scale which is much shorter than the shortest-lived nuclides, i.e., they must be available in much greater amounts. This is realized in another type of neutron-induced process which involves an incomparable faster sequence, the so-called rapid process (r-process). The latter requires an extremely neutron-rich environment, such as that in the explosion of a thermonuclear weapon or in a massive star.

NUCLEOSYNTHESES IN SUPERNOVA EXPLOSIONS

A supernova is a star that is heavier than the Sun. When it runs out of hydrogen fuel, it contracts under its own gravitational force and implodes. The implosion is accompanied by an instantaneous, ultra intense burst of neutrons when the outer layers collapse. The extremely high pressure inside the stellar core supplies a further neutron flash by converting the protons into neutrons. The burst lasts for a few seconds and provides an enormous number of neutrons. The target nucleus captures successively dozens of neutrons within a minute fraction of a second and the gap appearing in the s-process due to short-lived nuclides is overcome by the r-process. This allows the nucleosynthesis of all heavier chemical elements including the hypothetical superheavy elements.

The artificial nucleosynthesis of einsteinium (Z = 99) and fermium (Z = 100) by the r-process was demonstrated in 1952 in the explosion of a thermonuclear weapon.

In nature, it is most likely that only supernova explosions can provide the necessary conditions for the r-process. There are two types of supernova explosions. One of these (called type I) represents the final evolutionary stage of an old star of relatively moderate mass, corresponding to 1.2 to 1.5 solar masses. The entire star disintegrates in a giant thermonuclear explosion within a few seconds. The temperatures reached in various layers of the star range from 10^9 to 10^{10} degrees kelvin. The second type of supernova explosion (type II) occurs only in stars with masses at least eight to ten times that of the Sun. The core of these stars consists of onion-like layers which represent different stages of the star's life. A computer model suggests that these layers consume a progressive series of nuclear fuels: in the outer layers hydrogen is converted into helium; in the next, helium into carbon and oxygen; and, closer to the stellar core, carbon and oxygen give rise to neon, magnesium, and silicon.

After the last three nuclides are fused into iron and nickel in the core, gravitational contraction rises the temperature to about 5×10^8 degrees kelvin, causing a photodisintegration of iron and nickel into neutrons and ^4He nuclei, and accelerates the gravitational collapse. ^4He is dissociated into its nucleons (two protons and two neutrons), and the protons capture electrons and are converted to neutrons. Within 1 second, the entire stellar core collapses into an extraordinarily compact mass of neutrons with a density of 10^{14} grams per cubic centimeter. A neutron star is born, and its formation is accompanied by the explosive ejection of the star's outer layers into the interstellar medium.

Supernova explosions mark the death of a less common type of star and they are rather rare on the scale of a human lifetime. The frequency in the Milky Way is one every 20 to 50 years. This estimate is based on records of historically observed events and the observation of supernova explosions outside the Milky Way. The flash of the explosion, intrinsically brighter by several orders of magnitude than the Sun, fades away within a few days or weeks, while the dust cloud expands during a period of several hundred years and may reach a diameter of a few light-years. One of the best known of these occurrences is represented by the Crab nebula, the relic of the supernova explosion observed by Chinese astronomers in the year 1054.

Early in 1987, a "nearby" supernova explosion provided the first occasion for the present generation to observe such an event. It occurred in the galaxy nearest to the Milky Way, the Large Magellanic Cloud, which is at a distance of 170,000 light-years from the Earth. The neutrinos released during collapse of the star core have been detected (Chapter 2).

The remnants of supernovas are the stars known as pulsars, or pulsating stars. These are rapidly spinning bodies which emit pulses of energy in the radiowave regions of the electromagnetic spectrum. The pulses are perceived regularly like signals from a lighthouse; for example, the remnant of the "Chinese" supernova twinkles 30 times per second.

The pulsars are thought to be examples of neutron stars. They are very small and are so compressed that enormous pressures produce the inverse of β-decay in the core: electrons and protons are forced to combine to form neutrons.

In a more profound degree of collapse, the residue of a supernova would condense into a black hole. It is a hypothetical region of space possessing a gravitational field so intense that no matter or radiation can escape from it, although both could be continuously captured from the outside. For this reason they are invisible and their supposed existence relies on observations of various indirect effects.

In contrast with larger stellar bodies which collapse to neutron stars, or with massive stars which end as black holes, the deaths of smaller stars like our Sun are less dramatic. Once they reach the dimension of a red giant, they smoothly lose weight and volume by ejecting stellar material. The residue becomes a white dwarf, a star which has the size of the Earth but a density of 10^9 grams per cubic centimeter. It cools continuously for 1000 million years until it becomes a black dwarf.

The supernova explosion releases an enormous amount of energy: in the form of visible light alone, this can be as high as 10^{42} joules. The release occurs within 1 second and corresponds to the amount of light that the Sun emits over a period of 1000 million years.

The exploding supernovas also eject large amounts of matter and supply interstellar space with dust and chemical elements heavier than iron. Abundant amounts of radioactive materials are liberated and act as localized sources of energy in the interstellar dust and gas. The importance of these sources does not lie in their contribution to the total amount of energy liberated in the explosion, but rather in the specific and effective chemical action of radiation they emit (Chapter 6).

The supernova explosion also triggers a shock wave in the interstellar medium, heating the gas and causing turbulence. This can initiate the accretion process in the gas cloud. It may lead to formation of a new star, as was most likely the case for the presolar nebula and the birth of the Sun and Solar System (Chapter 7).

The ejected debris of supernova contains enormous amounts of hydrogen, helium, and many heavier elements in a more or less ionized state. The particles acquire increased kinetic energy in the explosion and, together with simultaneously released electromagnetic radiation, constitute the cosmic rays.

COSMIC RAYS

A fine rain of energetic particles and photons falls upon the Earth from outer space at nearly the speed of light, at all times and from all directions. It bathes our present planet just as it did the early Earth. All celestial bodies are exposed to this cosmic radiation.

Cosmic rays consist of an ionizing radiation like that from radioactive substances. However, there are significant differences with respect to their intensities and energies. Whereas the intensity of cosmic radiation is relatively weak, corresponding to a few species per square centimeter per second, 1 gram of a radioactive substance like radium emits thousands of millions of particles per second. On the other hand, the radiation energies from radioactive substances (several megaelectronvolts at the most) are weak when compared to the multimegaelectronvolt levels of cosmic rays. A single cosmic ray with very high energy, for example 1.5×10^{20} electronvolts, carries 25 joules, which is sufficient to raise a 1 kilogram weight to a height of 2.5 meters. And this energy is associated with a single body (an atomic nucleus) only 10^{-15} meter across.

The terrestrial atmosphere provides an efficient shield against cosmic radiation. A flux of approximately 20 particles per square centimeter per second at the upper limit of the atmosphere is reduced at sea level to about one particle per square centimeter per second.

Persons at the Earth's surface are not aware of the presence of cosmic rays, but cosmonauts on their way to the Moon were. In the Apollo spacecraft, in the absence of atmospheric protection, the cosmonauts experienced the effects of cosmic rays when attempting to sleep. Whenever an energetic particle penetrated the vessel and occasionally struck the retina of the closed eye, the dozing cosmonaut saw a tiny flash of light.

It is generally considered that above an altitude of 25 kilometers primary cosmic radiation is predominant; this form is also called galactic because most of it originates within our galaxy. When the primary cosmic rays strike the atoms and molecules of the atmosphere, they generate showers of subatomic particles which constitute secondary cosmic radiation. Most of this is absorbed in the atmosphere, or decays as short-lived radioactive entities before reaching the Earth's surface.

Primary Cosmic Radiation

This form provides samples of matter from outside the Solar System. Most of this material comes from within the Milky Way, but some is probably of extragalactic origin.

Cosmic rays lose energy in reactions that take place along their pathways. These reactions have definite lifetimes which depend on the type and energy of radiation. The galactic cosmic rays which are observed at present must have been produced relatively recently on the cosmic time scale, i.e., not more than 1000 million years ago. Radiations from the more distant galaxies do not reach the Earth.

The flux of primary cosmic rays consists mainly of protons. The fluxes of electrons and helium nuclei are each about 5 percent of that of the protons, while the γ rays are less abundant. The nuclei of all atoms heavier than helium follow with an abundance distribution matching roughly that of elements in the universe, and represent a fraction of 1 percent.

The energies of cosmic rays vary within a large range of up to 10^{11} gigaelectronvolts (10^{20} electronvolts); their fluxes decrease rapidly with increasing energies. The most important constituents of cosmic radiation are protons of about 2 gigaelectronvolts (2×10^9 electronvolts); the flux of 2000 gigaelectronvolts of protons is 10 million times smaller (Figure 22).

Little is known of the origin of cosmic rays or of the processes by which they are generated and accelerated to velocities approaching that of light. Potential sources of cosmic rays are supernova explosions, pulsars, and the shock-wave acceleration of matter in the interstellar medium.

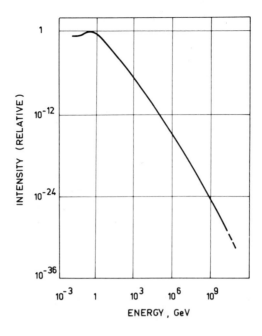

FIGURE 22.   Flux of protons in the primary cosmic radiation.

There are arguments that supernova explosions may directly accelerate the ejected particles up to 10^8 gigaelectronvolts, but it is more probable that most of the cosmic rays initially have moderate energies. The cosmic material, whatever its origin, is mixed with interstellar dust and gas and is gradually accelerated as it travels through space. The acceleration is due to cosmic magnetic fields that act like a gigantic accelerator.

Most of the particles never gain enough energy to escape from the galaxy, either because of the insufficient strength of the accelerating electromagnetic field or because the particles lose energy in collisions with other particles along their pathways.

The extragalactic cosmic rays are considered to be the particles which were accelerated sufficiently to escape the magnetic field of a galaxy. Particles of very high energy (10^9 gigaelectronvolts or more) certainly belong to this category, since no magnetic field would be strong enough to retain them within the galactic disk.

Studies of the most energetic cosmic rays are difficult because of the low intensities involved. On the average, 1 square meter of the Earth's surface is struck only once in a century by a proton having 10^8 gigaelectronvolts or more. Observation of such rare visitors is facilitated by the large-area detectors, which are made from a plastic scintillator, a material that emits light when an energetic particle passes through it. Arrays of scintillation detectors which cover several square kilometers have been constructed. Every array consists of dozens of detectors, each of which is several square meters in area. Twenty years after the first observation of a cosmic ray of 10^{11} gigaelectronvolts, the total world record of such events was only about 10. However, new techniques are emerging and a more complete collection of information is already within reach.

Solar Wind

Occasionally the Sun ejects protons in enormous numbers, sometimes several orders of magnitude higher than that of the primary flux of cosmic rays. This phenomenon occurs during the "flares", when high temperature bursts are seen as bright areas on the Sun's surface. Although the flare is optically visible for less than 1 hour, the increased flux of energetic radiation is observed in the vicinity of the Earth for more than 48 hours.

The frequency of flares corresponds to the 11-year cycle of sunspot activity. The ionizing radiation associated with solar flares may present a potential danger for cosmonauts, particularly during their activity outside the spacecraft, but it is relatively harmless for the population on Earth (Chapter 8).

The maximum energy of a solar ray is of the order of 0.1 gigaelectronvolt, or even higher in the case of exceptionally strong flares. On the average, however, the energies of typical solar protons are lower and are found mainly in the kiloelectronvolt range. These particles cannot penetrate the atmosphere and therefore do not reach the surface of the Earth. Part of the solar wind remains trapped in the Earth's magnetic field and forms the outer Van Allen radiation belt. Other, more energetic solar protons, penetrate the upper atmosphere and accumulate in narrow zones in the region of the Earth's magnetic poles, producing the well-known auroral displays.

Secondary Cosmic Radiation

In collisions with the constituent nuclei of the atmosphere, the primary cosmic rays produce a large number of secondary particles. The abundance of this secondary cosmic radiation depends on the flux and energy of the primary cosmic rays. For example, an energetic particle (say 6×10^9 gigaelectronvolts) initiates an enormous shower of secondary particles, which builds up to 10,000 million particles by the time it reaches sea level. At the core of each cosmic shower, there is a nuclear cascade involving a chain of nuclear reactions. This begins with a collision between two atomic nuclei, involving the primary cosmic ray particle and the nucleus of the constituent atom of the atmosphere. Many, but not all, of the secondary cosmic rays produced by such a collision subsequently undergo similar collisions along the axis of the cascade, producing secondaries of their own and so forth.

A large variety of elementary particles appears in secondary cosmic radiation. Before physicists had powerful accelerators at their disposal, cosmic rays provided the main source of elementary particles and enabled the discoveries of the positron and muon in the 1930s, several kinds of mesons in the late 1940s and various kinds of baryons in the early 1950s. Such studies gave birth to a new branch of physics called particle or high-energy physics.

Cosmic Rays and Cosmology

For particle physicists, the whole universe consists of a giant "accelerator". The properties of the universe provide information on physical laws at extremely high energies, and give hints on how to construct a unified theory of matter and energy. For cosmologists, such studies provide ways of visualizing how the universe got to be in the state we see it today, why it expands the way it does, and whether this expansion will continue forever.

The development of high-energy accelerators has enabled the attainment of energies rivaling those on the cosmic scale. From the time of the first machine named the Cosmotron, which produced 3 gigaelectronvolts (1952, U.S.), up to the present day with its thousand gigaelectronvolt accelerators, many objectives have been attained. Yet, it seems that particle physicists and cosmologists are finding a renewed interest in cosmic rays. The reason is that only the latter can offer the energies which are required for further progress in the knowledge of forces, elementary particles, and early stages in the evolution of the universe. These energies are far beyond the reach of accelerators of the present and foreseeable future.

ANTIMATTER

We ourselves and the world around us, together with the celestial bodies, are composed of matter in which negative electrons orbit around nuclei consisting of positively charged protons and neutral nucleons, the neutrons.

Theoretical physics predicts that for each charged elementary particle there should exist an antiparticle, a species with an opposite charge but with all other properties strictly identical. The simplest and most readily available antiparticle is the positron. It has the same mass as the electron and an electric charge of equal magnitude but of opposite sign.

Positrons are emitted in the decay of certain neutron-deficient radionuclides. When they collide with electrons, both particles are annihilated and their masses converted into electromagnetic radiation whose energy, according to the mass-energy relation, is twice the mass of an electron (2×0.511 megaelectronvolts). The energy is manifested as two photons created in the annihilation process.

The antiparticles of most of the known elementary particles have been discovered.

In the Universe

What was the situation when the universe was born? If, at that time, there existed a precisely equal number of particles and antiparticles, they would all have undergone annihilation as the temperature dropped below 1000 million degrees and nothing would have remained but the annihilation radiation. If that had been the case, there would be no matter in the universe and, of course, we would not be here!

Since matter exists, there must have been a slight excess of matter with respect to antimatter. According to the grand unified theories (GUT), this preference of the universe for matter over antimatter is due to the fact that the rate of baryon-antibaryon annihilation exceeded the rate of their regeneration. About 100 microseconds after the Big Bang, after all the antibaryons had been consumed in annihilation reactions, only 1 odd baryon out of 10,000 million was left over. It must have been this tiny remnant of primordial matter that gave rise to the visible universe.

Theory suggests that the subtle manifestations of the original matter-antimatter asymmetry should still be apparent. Since any contact between antimatter and matter results in a violent annihilation (and production of radiation), one fact is certain: if the asymmetry exists, matter and antimatter should be well separated. As a consequence — and science fiction dotes on the subject — there should somewhere be antigalaxies in the universe, i.e., complete galaxies composed of antistars, antiplanets, and, — why not, — inhabited by "antipeople".

The cosmic ray showers have long been known to contain antimatter, in particular the antielectron and antiproton. But in order to be certain that these also represent the primordial components of the universe, the search has to be carried out beyond the Earth, in the realm of primary cosmic radiation. Results of such studies reveal only a scanty presence of antiprotons in the primary cosmic proton rays. The former could conceivably also originate during the interaction of primary cosmic radiation with interstellar gas and dust, and thus a clear-cut answer is still lacking.

Artificial

Since man-made antiparticles are available, why should we not also produce antimatter in the laboratory? For example, beams of antiprotons (p⁻) and positrons (e⁺) could be merged as follows:

$$p^- + e^+ \rightarrow \text{antihydrogen (anti-H)} + \text{photon}$$

The basic reaction involved would be the so-called "spontaneous radiative capture" of a positron (e⁺) by an antiproton (p⁻). This is a relatively rare process and it requires stimulation; the use of a laser beam for this purpose has been suggested, but its feasibility still has to be demonstrated. Another problem that presently remains to be solved is that the antiparticle beam intensities are too low, especially in the case of positrons.

There seem to be fewer obstacles in connection with the utilization of energy liberated during the annihilation of antiparticles with matter. A basic requirement is the trapping of antiparticles under special conditions so that, once a sufficient amount of antiparticles has been accumulated, the trap can be used elsewhere as an antiparticle source. At present, such traps are technically feasible.

Although at present only an extremely small number of antiprotons can be collected in antiproton traps, and these last for several minutes at the most, public opinion has been alarmed by the potential uses of antimatter for military purposes. Production of about 10 milligrams (0.01 gram) per year of antimatter would be considered by the U.S. Air Force to be "interesting" on a military scale. The energy released in the annihilation of 1 milligram of antimatter with matter would be equivalent to that produced by 44 tons of TNT. Less than a microgram (10^{-6} gram) of antiprotons is estimated to be sufficient to trigger a thermonuclear explosion or pump a powerful X-ray laser. A burst due to the annihilation of a few nanograms (10^{-9} gram) of antiprotons would be adequate for discrimination between active nuclear warheads and decoys in outer space.

Antimatter would not only be the most "portable" of high explosives, but also a convenient source of muons. Each antiproton annihilation produces an average of three muons. These could be used to catalyze the fusion process in a deuterium-tritium mixture (Chapter 10), thus providing a particularly attractive solution for a low-weight nuclear fusion reactor for spacecraft and orbital stations.

Various concepts of weapons based on antimatter are being worked out and a new generation of nuclear weapons seems to be within the reach of scientific achievement and it is not surprising that calls are appearing for an immediate ban on research related to antimatter.

The giant particle accelerators used now to make antimatter are inefficient and expensive. The antimatter is produced only in quantities measured in femtograms (10^{-15} grams). The cost is as much as 1,000,000 million dollars (10^{12} dollars) per microgram.

DARK MATTER

Contrary to popular belief, the interest of astronomers and physicists in the "material content of the universe" is not so much directed towards the bright stars and galaxies, but rather to the unseen dark matter and the so-called missing mass. The present state of knowledge indicates that dark matter is predominant in the mass content of the universe.

Dark matter is not discernable with respect to any part of the electromagnetic spectrum (quite apart from the "visible" region), yet its gravitational effects on normal matter can be observed. The rotation of spiral galaxies such as the Milky Way can be explained only if 90 percent of the matter is invisible at any given wavelength of the electromagnetic spectrum extending from γ-rays to radio waves. Similarly, the motions of clusters of galaxies imply gravitational effects that are much larger than those expected from the known mass. Various analyses and determinations suggest that the "visible" celestial bodies account for only 10 percent of the total material content of the universe.

Another important conclusion is that this dark mass exists in a form which is unknown on Earth or in the "visible" universe; it is not composed of baryons (protons and neutrons) and leptons (electrons).

Particle physicists are attempting to assist cosmologists by suggesting the existence of particles that might be remnants from the Big Bang. For some time, neutrinos were considered as a likely candidate for the missing mass because of their abundance. Theoretically, the neutrino would appear to be the only known particle existing in sufficient amounts in nature to constitute dark matter. However, there is a problem with respect to mass. If the neutrinos' rest mass were 20 to 30 electronvolts, as some set of laboratory measurements has indicated, then the total mass of neutrinos would definitely account for the observed gravitational effects and the problem of the nature of dark matter would be solved. However, other measurements suggest that the mass of the neutrino must be smaller than 20 electronvolts if, indeed, the particle has any mass at all.

Some important information might have been provided by the burst of neutrinos from the recent supernova explosion of 1987. This lasted only a few seconds, and a total of 19 neutrinos was detected independently in Japan and in the United States. The measurements made in Japan seem to support the idea that the neutrino's mass actually is less than 15 electronvolts, i.e., that they are not the constituent of dark matter as was previously believed. These measurements were performed with the most sensitive detector of its kind in the world, consisting of 2140 tons of purified water and an extremely delicate system for radiation measurement. The equipment was primarily designed for observation and recording of the radioactive decay of protons, but it is also suitable for detection of neutrinos.

In the event that the neutrino has insufficient mass to be considered as a possible constituent of dark matter, cosmologists and particle physicists will have to look for another candidate. A proposed substitute is the axion, a hypothetical particle which in some respects resembles the neutrino. Its mass would be small compared to that of the electron and unlikely to exceed 25 eV. The probability of the interaction of axions with other particles would be very low, so that direct experimental observation would be precluded. Axions should be abundantly produced in weak nuclear interactions such as those responsible for β-decay, particularly in the interior regions of stars. To some degree, axions should be related to neutrinos as muons are to electrons, i.e., they should represent a more massive version of the same species. To what extent the existence of axions (if proven) would successfully account for the present estimates of the missing mass in a gravitationally closed universe will depend on their number and individual mass.

In addition to promoting advances in cosmology, studies of the problem of dark matter are an effective stimulus to work in theoretical and experimental particle physics.

Further Reading

Weinberg, S., *The First Three Minutes* , Bantam Books, New York, 1977. [The evolution of the universe, presented clearly and with scientific accuracy. Suitable even for those who are not at home in either mathematics or physics.]

Close, F., Marten, M., and Sutton, C., *The Particle Explosion*, Oxford University Press, New York, 1987. [Provides elementary knowledge on radioactivity and the structure of the atom and describes the discoveries of elementary particles. An account is given of sophisticated equipment which is used in high-energy physics. Furnishes a detailed survey of the major particles and groups of particles. Characterized by clear presentation and excellent illustrations.]

Wolfendale, A.W., Ed., *Progress in Cosmology*, D. Reidel, Dodrecht, 1982. [Proceedings of a scientific meeting on cosmology, particle physics, and cosmic rays. A useful state of the art at the beginning of the 1980s.]

Sagan, C., *Cosmos*, Random House, New York, 1980. [An excellent presentation of various aspects of the universe. Clearly written and extremely well illustrated.]

Galant, R.A., *Our Universe*, National Geographic Society, Washington, D.C., 1980. [The National Geographic picture atlas of the universe, with emphasis on the Sun and planets. Concise text and rich illustrations.]

Field, B.G., Vershuur, G.L., and Ponnamperuma, C., *Cosmic Evolution: An Introduction to Astronomy*, Houghton Mifflin, Boston, 1978. [This book is intended as a text for a course to be taught by instructors having a knowledge in astronomy. A knowledge of high school mathematics is sufficient. It may also be of help to all those who are interested in the astronomical aspects of the evolution of the universe, from the Big Bang to the current stage in which intelligent life has emerged.]

We have seen that interstellar space is both the burial ground for remnants of stellar systems which have expired and a place where stars and planets are born. Interstellar matter is reprocessed in the course of successive generations of stars in violent events at extremely high pressures and temperatures.

An essentially different type of chemistry is considered in this chapter. It concerns the rarefied mixture of gas and dust between the stars, which represents several percent of the mass of a galaxy. This chemistry takes place at extremely low densities of matter and low temperatures. Measuring equipment set up on the Earth and in satellites is presently supplying an ever-increasing store of information on the presence of fairly complex molecules, and about 100 species comprising ions and free radicals have been reported. The latter are produced to a large extent by ionizing radiation and are mainly responsible for the interstellar syntheses.

Radiation and radioactivity must also be taken into account in the investigation of comets and meteorites. Comets are bodies composed of ice and dust with a diameter up to a few kilometers. They travel through the confines of the Solar System as a cometary cloud until some of them approach the inner Solar System. Meteorites are bodies, with a size ranging from micrometers to several meters, which reach the Earth's's surface after revolving around the Sun. As relics of events that took place almost 5000 million years ago, they provide us with information from the era prior to the formation of the Solar System from presolar nebulas.

The spectacular close encounter with Halley's comet in 1986 provided a wealth of data. They suggest that prolonged exposures to cosmic rays have led to important chemical changes in primordial cometary material. Laboratory experiments show that radiation-induced synthesis can lead to significant amounts of fairly complex organic molecules. In the early period of the Solar System when cometary impacts occurred, organic compounds may have reached the primitive Earth. Cosmic rays also induce the formation of radionuclides, which are of particular help for the radioactive dating of meteorites and which provide reliable information on the time scale of various events in the life of the Solar System.

6

The Interstellar and Interplanetary Medium

BETWEEN THE STARS

Space

The average density of interstellar space is very low, corresponding to about only one atom or less per cubic centimeter, and the temperature may be as low as 3 degrees kelvin (–270 degrees celsius). However, these levels of density and temperature are not evenly distributed and various regions in space may have drastically different characteristics.

An imaginary route from a star out toward interstellar space would pass first through the stellar atmosphere, whose composition would depend on the type of star and its surface temperature. For a younger star the latter is about 10,000 degrees kelvin and the density lies between 100 and 1000 ionic species per cubic centimeter. These species comprise mainly hydrogen ions but also those of carbon, nitrogen and oxygen. Some cooler stars eject their matter and create a nebula of dust and gas in the surrounding medium. Temperatures lie between 100 and 1000 degrees kelvin, and the nebula may also contain silicon or carbon if the stars are rich in these materials. Farther away from the stars, the interstellar medium consists of a dilute background gas with 0.1 species per cubic centimeter, and in which hydrogen atoms and carbon ions dominate.

However, there are regions in interstellar space in which a significant amount of matter is condensed, for example, in interstellar clouds, in which the mass often exceeds thousands of solar masses and where many different types of molecules can exist.

There are two important types of interstellar clouds. One of these is the diffuse type, in which the mass is rarefied and optically transparent, with a density not exceeding several hundred atoms per cubic centimeter. The gas temperatures lie between 50 and 100 degrees kelvin, with a value of about 10 degrees kelvin for the dust grains. The main constituent is hydrogen, of which the atomic form is predominant, the rest of it consisting of H_2 molecules. Other species include carbon monoxide (CO) and formaldehyde (HCHO).

Most of the chemical compounds known at the present time have been observed in the dense interstellar clouds, opaque masses of molecular hydrogen in which the density is about 10,000 molecules per cubic centimeter and the temperature 5 to 10 degrees kelvin. Inside the cloud cores the temperatures are at least 100 degrees kelvin and the densities higher than 1 million H_2 molecules per cubic centimeter.

Gas

For probing the chemical content of interstellar space, scientists combine radioastronomy with microwave spectroscopy, a technique which is routinely used by chemists for the accurate determination of molecular structures. Over 100 species have been reliably identified, including molecules, ions, free radicals and their isotopic varieties, i. e., entities which have the same structure but differ in isotopic composition (for example, replacement of hydrogen by deuterium).

A dense molecular cloud designated as Sagittarius B2 is located at the center of the Milky Way. Its mass is equivalent to 3 million solar masses. More chemical species have been detected in this cloud than in any other astronomical source.

The list of interstellar species shows that many, including the most abundant, are of simple molecules of the same kind as found on Earth, e.g., water (H_2O), carbon monoxide (CO), or formaldehyde (HCHO). Some are known only in the laboratory, such as the hydroxyl radical (OH); they are chemically reactive and under routine experimental conditions can exist only a small fraction of a second. Others are unknown on Earth and still have not been produced even in the most sophisticated experiments. Such is the case for a compound called cyanopolyyne, which consists of hydrogen, carbon and nitrogen in a chain of 13 atoms: HCCCCCCCCCCCN. This compound has been detected in the proximity of a star that is at a distance of 660 light-years-from the Earth.

The present inventory of interstellar compounds reveals only the "tip of an iceberg". It is highly probable that numerous other compounds known on Earth exist in the interstellar medium. Moreover, because of the very unusual environmental conditions which exist in space, many other compounds unknown on our planet would be expected to form there. Interstellar space is a natural laboratory where chemical reactions take place in a vast vessel without walls, at extremely low pressures and temperatures, whereby even the most reactive species can subsist long enough to participate in reactions which are unknown on Earth. By far the most abundant chemical compound in interstellar space is molecular hydrogen. For each 1000 million (10^9) H_2 molecules there exist 10,000 of carbon monoxide (CO), 100 hydroxyl radicals (OH), 10 cyano radicals (CN), ammonia (NH_3) or formaldehyde (HCHO), and 1 of hydrogen cyanide (HCN), sulfur dioxide (SO_2), methanol (CH_3OH) and ethanol (CH_3CH_2OH).

Various sources of energy are available for chemistry in interstellar space. In the optically thin masses of diffuse interstellar clouds the light of bright stars is an important source. This radiation covers not only the visible part of light, but also the more energetic part of the electromagnetic spectrum. Stellar winds, which are the fluxes of charged particles emitted from the surfaces of stars, also contribute to the formation of ions and other chemically reactive species. In the dark molecular clouds, the role of these sources is limited because of an increased absorption of light and stellar wind in the dust and gas. In this case, sources such as shock waves or cosmic rays are more effective.

A shock wave can raise the local temperature up to several thousand degrees. It occurs when the constituent of the gas moves at supersonic speed. Interstellar gas is stirred up by many kinds of violent events (e.g., a supernova explosion) which can cause certain gas layers to attain such speeds. The ensuing temperature increase leads to the formation of excited and ionized species, and consequently of ions and free radicals which initiate chemical reactions. The density of dark molecular clouds is not an obstacle to cosmic rays, an ionizing radiation that consists of energetic electromagnetic waves and charged particles.

Ionization of the most abundant interstellar species, the atom and molecule of hydrogen, is of great importance for chemistry in space. This process leads to formation of H^+ and H_2^+ whose behavior is reasonably well understood from laboratory experiments and explains many interstellar reactions.

Dust

Interstellar matter consists of about 70 percent by weight of hydrogen and 28 percent by weight of helium, other chemical elements representing the remaining 2 percent. Some of the heavier elements like carbon, oxygen, magnesium, iron, or silicon are depleted in interstellar gas and are assumed to be incorporated in what is called the interstellar dust grains. It is estimated that the total mass of interstellar dust is 100 times lower than the mass of gaseous hydrogen. Nonetheless, dust may have a preponderant role in interstellar chemistry.

No specimens of interstellar dust grains have as yet been gathered for laboratory examination, but their existence is supported by various observations.

A significant, although indirect argument in their favor is that the existence of interstellar dust must be assumed in order to explain the origin of molecular hydrogen (H_2). The high rate of formation of this molecule cannot be explained by presently known reactions in the gaseous state without the participation of a solid surface.

More direct supporting evidence stems from well-established facts such as the observed ejection of stellar material from surfaces of certain types of stars, and the nebulosity of stars in or near an interstellar cloud. Further, the existence of diffuse visible and ultraviolet light components in interstellar space can be explained only by the scattering of light due to dust particles.

The only experimental approach to interstellar dust grains is the examination of light from a star which reaches the Earth after passing through a "dusty" region of the sky. It is known from laboratory experience that one way of influencing the propagation of light is to interpose along its path particles whose diameters are of the same order of magnitude as the wavelength. The size of absorbing particles in astronomical observations is inferred from the way in which the star's light is affected.

Studies show that the optical spectra are affected in several regions: in the visible, in a manner typical for particles of 0.3 micrometer; in the ultraviolet, which corresponds to 0.1 micrometer and less, and in the infrared, where the particles should be at least ten times larger. Broadly speaking, the sizes of cosmic dust grains are of the order of 1 micrometer and are comparable to those of particles in cigarette smoke.

Although dimensions alone do not give a clue to the problem of chemical composition of dust grains, they help in getting some insight into it. It is possible to reproduce in the laboratory the observed dependence between wavelength and particle size. For this purpose, mixtures of silicate and graphite particles of submicrometer dimensions are used in proportions suggested by the observed composition of interstellar gas. The experiments support the idea that amorphous carbon, silicon, and oxides, such as those of iron and magnesium, may be the main internal constituents of dust grains.

The cores of such dust grains are most likely the particles ejected from the stellar surface and propelled into space by the pressure of radiation emitted by the star. When their distance is sufficient to allow cooling down to about 10 degrees kelvin, these particles become seedlings for the growth of interstellar grains. Their size gradually increases by fixation of atoms and molecules which stick and freeze on to the surface. It is certain that the actual mechanism of growth is more complex than is implied by mechanical sticking and freezing alone. Some kind of chemistry may also be involved, for example, action of ionizing radiation and free radicals generated in the grain structure during growth.

The formation of a molecule by interstellar dust-grain surface chemistry is more complex than a synthetic process in the gas phase. Many aspects have to be taken into account. One is the adsorption on the grain surface of a free radical or an atom from the gas phase, its migration and reaction with another species, and finally the ejection or evaporation of the product molecule back into the gaseous environment. The role of temperature and physical and chemical properties of the grain surface is far from being understood and, despite significant achievements, approaches to interstellar grain-surface chemistry still remain highly speculative.

In laboratory experiments one can simulate the effects on ice of stellar wind, which consists of particles with energies in the kiloelectronvolt-megaelectronvolt range, and the action of energetic cosmic ray particles which are mainly in the megaelectronvolt-gigaelectronvolt range. The phenomena investigated concern erosion and chemical alterations.

In the erosion experiments, thin ice films (less than 1 micrometer) are bombarded with accelerated ions (usually those of hydrogen) with energies between 10 kiloelectronvolts and 2 megaelectronvolts. Analysis is carried out on the material ejected from the film during bombardment and on the residue following irradiation and thawing of the sample. In conjunction with carbon-containing materials, such experiments have shown that organic syntheses can take place; for example, formaldehyde can be identified in an irradiated mixture of water and carbon dioxide. The erosion of ice due to bombardment by low-energy ions is a slow process which, in interstellar space, is estimated to be of the order of 1 nanometer per year. However, the small dimensions of interstellar ice grains together with the available time spans of hundreds of millions of years for erosion make the phenomenon important not only for the loss of the icy mantle on the dust grain, but also for the supply of interstellar molecules in cosmic space.

Investigations of chemical alteration of ice by radiation usually consist in bombarding simple molecules like methane (CH_4) at liquid helium temperature, with protons in the kiloelectronvolt-megaelectronvolt range. It has been observed that solid complex organic materials are formed in an amount which increases with the number of impinging ions. With methane, the residue is an amorphous, fluffy material of low density and the infrared spectra indicate that the substances are polymer-like. Continuing treatment leads to an amorphous, carboniferous material which loses hydrogen and whose density increases as a function of radiation dose. Such a polymer may be the glue which bonds the submicron silicate particles to form micrometer-sized aggregates.

A technique called ion implantation, well known in solid-state physics, also provides insight into possible radiation-induced chemical changes in interstellar dust. The method consists in injecting (implanting) the accelerated ions, which have higher energies than those of a chemical bond. An example is the implantation of accelerated carbon ions in silicate grains. Spectroscopic examination of the irradiated samples reveals *in situ* formation of CO_2 or CO, depending on irradiation conditions. The experiments offered an explanation for the unexpected observation of these gases in lunar dust grains.

When a charged particle interacts with the frozen material, "hot" atoms are abundantly created along the path of the collision cascade. The chemistry based on the effect of activated (hot) atoms is presented in some detail in Chapter 4. Certain findings relevant to hot atom chemistry in the interstellar medium, in particular with biogenic atoms like carbon and nitrogen, are worth mentioning. The irradiation of ice forms (frozen water, ammonia, or methane) shows that reactions of hot carbon and nitrogen atoms form a wealth of compounds. These include carbon monoxide, carbon dioxide, formaldehyde, formic acid, or methanol, in addition to some other simple compounds like methylamine, cyanamide, formamidine, and guanidine. Larger molecules are also produced, such as hydrocarbons containing up to 13 carbon atoms.

Insight into radiation-induced reactions is complicated by the fact that the analyses must usually be performed after the sample is thawed, i.e., when additional (postirradiation) effects may be induced in the products. In principle, it is more convenient to use a nondestructive method such as spectroscopy, which allows *in situ* analysis of the thin film of ice while it is still under bombardment with low energy radiation. Nevertheless, this approach has a drawback: a rather large number of bombarding particles (10^{14} to 10^{17} ions per square centimeter) is needed to induce sufficient change for reliable measurement. As a result, the high radiation doses involved lead to complications in the interpretation of the process. It is obvious that, wherever possible, classical and *in situ* analyses should be judiciously combined.

Ultraviolet radiation has been used in experiments at 10 degrees kelvin with water ice containing carbon monoxide, ammonia, or methane. The infrared spectra recorded *in situ* showed characteristic functional groups of many simple compounds which were likewise produced in the experiments with charged particles. Such experiments have also provided information on the possibility that material is ejected from the dust grain mantle. Explosions were recorded cinematographically for the part of the mantle in which the temperature increase was of the order of 25 degrees kelvin. The explosion energy was presumably released during exothermic recombination reactions of free radicals. In order to raise the temperature of an interstellar grain from 10 degrees kelvin (temperature of the experiment) to 30 degrees kelvin, 1000 radical-radical reactions have to occur. The heat generated can liberate other radicals and cause them to diffuse through the rigid matrix until they encounter other radicals with which they react. In this way, additional energy is liberated, triggering a chain process which results in an explosion and ejection of matter.

Photolytic experiments provide useful complementary information on chemistry involving charged particles in a stellar wind. The action of low-energy electromagnetic radiation (such as ultraviolet light) produces reactive intermediates similar to those formed during interaction of charged particles with frozen states. However, the nature of these species and the pattern of their spatial distribution are less complex than those produced by charged particles.

In comparison with chemical processes initiated by ultraviolet light, radiation chemistry has a great advantage. Owing to the high penetrating power of cosmic rays, it can take place even at great distances from stars and in the dense molecular clouds where most of the changes involving interstellar dust occur. The radionuclides imbedded in the dust grain also provide localized sources of energy: the radiation emitted by radioactive decay produces free radicals *in situ* which promote chemical reactions inside the grain.

Down to Earth

An interstellar dust grain has a life cycle of 100 million years or more, during which it is transported from a diffuse to a dense molecular interstellar cloud. Eventually, if it does not become incorporated into a newborn star, it again enters a diffuse cloud.

A celestial body such as our planet continually accretes matter from space and it is likely that interstellar grains reach the Earth. Interstellar dust could have previously reached the surface of our planet by the impacts of comets, which contain the dust material of presolar nebula. Moreover, in its travels through space the Earth has probably traversed several dense interstellar clouds, especially during the early stage of its existence. The estimates of amounts accreted vary from 100 million to 10,000 million tons of interstellar dust.

It would be premature to claim, as has sometimes been stated both in the scientific literature and mass media, that interstellar dust contains complex biologically important compounds, or that it provided the molecular templates for the origin of life on the primitive Earth. The chemistry of interstellar dust grains is far from being fully understood and requires many more astronomical observations and laboratory experiments.

COMETS, VISITORS FROM THE OUTER SOLAR SYSTEM

Comets are among the most spectacular celestial phenomena which can be observed from the Earth. They can occasionally appear as an impressive giant beam of light, shining with a long tail in the night sky. Records of the appearance of comets extend over at least 3 millenia of man's history.

Relics of Presolar Nebula

About 1000 comets and their orbits have been recorded up to the present time. The British astronomer Edmund Halley (1656—1742) was the first to notice the regularity of cometary appearances. After studying reports on many comets, he concluded that one of them (which was to bear his name) returns at regular intervals of about 76 years. We know now that regular comets have periods varying from several years (like comet Enke, 3.3 years), to 10,000 years and more like comet Kohoutek, that was observed in

1974 and will not be back again for at least 200,000 years. There are also nonperiodical comets that visit us only once. Their orbits are elongated and inclined at an angle to the plane of the Solar System, in contrast with periodical comets which have elliptical orbits situated near the plane in which the Earth and other planets revolve.

The period of a comet is analogous to our planet's year, representing the time the comet requires to complete its orbit around the Sun. The long-period comets like Kohoutek have very eccentric orbits that take them in their course far beyond the outermost planets of the Solar System. The loss of material which occurs during the comet's passage close to the Sun, together with gravitational interactions with the major planets, can gradually transform them into short-period comets.

A comet becomes visible when it approaches the Sun and solar heat liberates volatile constituents from its nucleus, which is a lump measuring usually no more than 10 kilometers in section. The characteristic appearance of a comet results from fluorescence and the reflection of sunlight caused by the presence of gases and particles expelled from the nucleus.

The evaporation of material becomes more intense when the comet's distance from the Sun decreases. A diffuse, glowing envelope called the coma develops around the nucleus. The coma may extend for several thousand or even a million kilometers.

Part of the materials liberated from the comet nucleus streams away for a distance of up to 100 million kilometers or more. This is the cometary tail, which occasionally extends in a spectacular manner across the sky. The effect of pressure associated with light and solar wind always causes the tail to point away from the Sun; this is evidenced by the fact that the tail follows the comet's head during approach and precedes it when the comet recedes.

Observations of comets are not always an easy or rewarding task. The physical state of a comet and the nature of its observable physical-chemical processes depend on its distance from the Sun. A complete observation of the comet's passage must be carried out while it is discernible, i.e., within a period of days or weeks.

Every year two or three new, long-period comets are discovered, but the tails of most of them, like those of their older and better-known counterparts, are generally not spectacular. Moreover, the reappearance of a periodical comet can be predicted with reasonable accuracy in contrast with the visual extent of its tail; we witnessed an example of this during the passage of Halley's comet.

The absence of a visible comet tail is not particularly disturbing for scientists. Many techniques are now available for cometary observations which do not depend on visibility, e.g., ultraviolet and infrared spectroscopy or radioastronomy.

Spectroscopic observations from the Earth's surface have provided a great deal of data on comets, despite serious limitations imposed by the atmosphere which absorbs electromagnetic waves in various regions of the spectrum. Scientists have considered sending a spacecraft to a cometary nucleus in order to take samples of its material and bring them back for analysis. In a case such as Halley's comet, it would require a little more than a year to reach the comet and about 5 years to return to Earth. The achievement of automatic sampling and return could be realized on the basis of our present technical knowledge and the success of the close encounter of spacecrafts with Halley's comet in 1986 is encouraging. It was possible, on that occasion, to approach the comet and examine it from a relatively short distance, whereby photographs were made and various measurements using board instruments were performed.

A main reason for the present interest in cometary studies is the widely accepted belief that they contain the best preserved primordial material of presolar nebula. According to the Dutch astronomer Jahn H. Oort, as many as 100,000 million comets are circulating at the periphery of the Solar System, in a cold reservoir that remains from the time the Sun and the planets were formed some 4600 million years ago. Oort's cometary cloud has not yet been observed, but astronomers do not doubt that it exists at a distance of 30,000 to 100,000 astronomical units from the Sun. An astronomical unit represents the distance between the Sun and the Earth and corresponds to about 150 million kilometers. It is surmised that occasional disturbances, such as the gravitational tug of a distant passing star, could "snatch" comets and force them into a cigar-shaped orbit with Oort's cloud at one end and the Sun at the other. Such an event would offer scientists a fine opportunity for examining relics which date from the birth of the Solar System.

Observations which were made before 1986 had already accumulated sufficient data on the physical and chemical properties of comets to establish a cometary model. Data of potential interest regarding the comet's chemical composition were provided both from spectra recorded by instruments on Earth and from information furnished by satellites.

About 40 species were observed and assigned as cometary constituents. Only a few of them are stable molecules like carbon monoxide (CO), water (H_2O), hydrogen cyanide (HCN), and methylcyanide (CH_3CN). Other species include chemically reactive free-radicals (OH, CN, CH, NH, NH_2), radical ions (CH^+), ions (H_2O^+) or atoms (H, O, C, S). Metals (Fe, Co, Ni, Cu) were found only in the comas of comets which closely approached the Sun.

Records of solar light reflected by the dust tails of various comets suggest that silicates, as well as some organic materials, may be contained in the particles which are about 1 micrometer in size. The yellowish color of the dust tail is due to this reflection. The bluish color of what is called the gas or ion tail, on the other hand, has been assigned to H_2O^+ and CO^+, whose properties are well known from laboratory studies.

The organic material in the dust from comet Halley was analyzed by mass spectrometry on board of spacecrafts. The results suggested the presence of at least 20 varieties of organic molecules, most of which were highly unsaturated.

Even before the close encounter with Halley's comet, there were various indications that a compact body, or nucleus, must be concealed behind the visible head of the comet. A principal argument was that many comets can lose large amounts of material during each passage near the Sun and still persist throughout a considerable number of revolutions. Also, comets have been seen to split into separate components which move apart from each other with velocities of a few kilometers per hour. Five photographs of comet West that were taken by American astronomers during a period of 2 weeks in 1975 clearly show the successive stages of splitting into at least four components.

"Dirty Ice Ball"

In the early 1950s, the American astronomer F. Whipple had already proposed a hypothesis on the comet nucleus, known as the "dirty ice-ball". This hypothesis, which later received increasing support, considers that the basic cometary entity is a nucleus extending over several kilometers, composed of water and simple gases in the frozen state. The frozen volatile constituents are imbedded in water ice together with presolar dust grains. The ratio of volatile material to dust is tentatively assumed to be 1:1 and the density of the nucleus about 1. Thus, the total mass of a comet of radius 5 kilometers is about 500 million tons (5×10^{14} grams). According to this model, the Sun's heat liberates molecules from the surface of the nucleus, and the latter undergo various chemical changes in the gas phase, mainly as a result of photochemical reactions.

Assumptions on the density of the comet nucleus and its chemical composition were not entirely confirmed during the passage of Halley's comet in 1986 and uncertainties in these respects cannot be clarified until samples can be taken from various depths of the nucleus and analyzed in laboratories. The main obstacle arises from the fact that the present data are provided only from observations of the coma and tail. The compounds present in these parts probably consist mainly or even entirely of daughter products of the real constituents of the nucleus which were chemically altered (e.g., photolytically) while leaving the core. For example, the well-established species CO and CO_2^+ are only indicative of the existence of CO_2 as a parent compound, in the same way that observation of the CN radical implies the presence of HCN and CH_3CN molecules. In fact, the reactions of other compounds are also known to produce these species.

A branch of chemistry dealing with the rates of chemical reactions, called chemical kinetics, is useful in gaining insight into the true chemical composition of the nucleus. The kinetic approach is based on computer modeling of chemical processes which take place in the comet's coma and the comparison of predictions with the astronomical observation of the kinds and amounts of species in the coma. If there is no agreement between the computed and observed data, scientists can be certain that their assumptions about the nucleus are erroneous. However, discrepancies offer useful indications regarding the manner in which assumptions about the nature, abundances, and reactions of the parent molecules should be modified in the computer simulation program.

Computer modeling is primarily based on the photochemical activation of parent molecules and, in a first step, takes into account reactions of the daughter products of the first generation. The latter products undergo further transformation in chemical reactions and lead to the formation of a second (third, fourth, etc.) generation of products. The computer program handles the overall reaction scheme, which may consist of up to several hundred chemical equations; necessary data concerning rate constants of reactions and the nature of intermediate species are provided by laboratory studies. The simulation is repeated by introducing new data on reacting species and their reactions until the observed and predicted values fit reasonably well together. Whereas lack of satisfactory fitting provides a reliable argument that the input data are unsuitable, coherence in fitting only shows that the data used are possibly appropriate. The reason for this is that experimentally obtained data may fit with more than one process-scheme if incomplete or incorrect data for the rate constants of reaction intermediates are employed.

Some results of computer modeling of cometary chemistry are particularly worth mentioning. Modeling does not provide satis-

factory interpretations if only the four simple compounds originally suggested by Whipple (water, ammonia, carbon dioxide, and
methane) are adopted as the parent molecules. Better fitting is
achieved if, in addition, certain other interstellar molecules are
also taken as constituents of the nucleus, e.g., hydrogen cyanide
(HCN), methylcyanide (CH_3CN), methylamine (CH_3NH_2), formaldehyde (HCHO), methanol (CH_3OH), allene ($H_2C_3H_2$) and acetylene
(C_2H_2). This finding has led scientists to believe that comets may
have been formed either in the outer parts of the presolar nebula
or in a companion fragment of the parent interstellar cloud from
which the Solar System evolved (Figure 23).

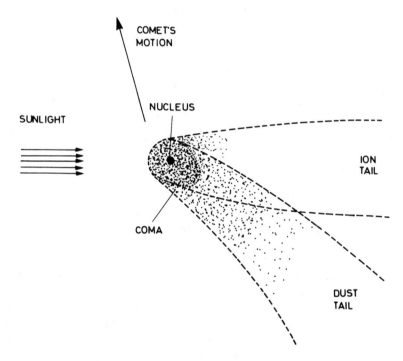

FIGURE 23. Principal features of a comet.

The First Close Encounter with a Comet

Five space probes, of which two were Soviet, two Japanese and
one European, passed near Halley's comet in March 1986. Numerous measurements were made while the spacecrafts were approaching the comet. In this hitherto unique undertaking, one of
the satellites (Giotto, launched by the European Space Agency)
took photographs and performed sophisticated measurements at
a distance of only 600 kilometers. Although the closest point of

approach lasted only about 10 minutes, it contributed most significantly to the already rich harvest of data which makes Halley the best known of all comets.

Photographs of the cometary nucleus show that it has the shape of a potato or a pear, with a cross section of 16 kilometers. The surface is smooth, but a hill of about 400 meters appears above the surrounding plain. There is also a kind of small "crater", 70 to 200 meters deep and about 1600 meters wide.

The total volume of the nucleus is estimated at about 400 cubic kilometers. Suggested values for its mass content vary from 50,000 to 130,000 million tons, but some scientists believe this may be a serious underestimate and that the real value may be up to four times higher. The average density of the cometary nucleus is considered to be between 0.3 and 1 gram per cubic centimeter.

The mass content has been estimated from the effects of cometary jets on the comet's orbit. These jets are due to ejection of gas and dust from active zones on the sunlit side of the nucleus. It is significant that, although half of the surface is illuminated by sunlight at any one time, only about 8 percent is active. At the time of Giotto's encounter, the rate of total mass loss was about 12 tons per second. The mass of the ejected gas was about three times that of the dust.

Identification of chemical compounds in the comet's coma was made by recording the optical spectra. The main neutral constituent of the gas in the internal part of the coma (less than 500 kilometers from the surface) is H_2O, followed by smaller amounts of CO_2, and the secondary species NH_2, OH, C_2, CH, CN, and NH. The majority of ions originate form water, viz., OH^+, H_2O^+, and H_3O^+.

The findings in the computer modeling of chemistry in the coma show that water should represent 84 percent of the volatile constituents of the nucleus, while the next most abundant components are likely formaldehyde and carbon dioxide (both present to about 3 percent). The nucleus probably contains carbon monoxide, nitrogen, carbon disulfide, and methane, to the extent of 1 to 2 percent, as well as smaller amounts of methylcyanide, acetylene, hydrogen cyanide, and methylamine.

The total mass of dust which came into contact with the Giotto spacecraft was estimated at several centigrams, and the mass of the particles varied from 1.1 milligram to 10^{-13} gram. It appears that the material ejected from the cometary nucleus contains rather large particles which consist of small ones held together by some sort of "glue". As they are blown away from the nucleus, they are heated by the sunlight, and the ice in the glue gradually melts and releases smaller particles. The analyses of particles with sizes ranging from 5 to 0.02 micrometer revealed that they were of two different types. The first contained considerable amounts of carbon, oxygen, sodium, magnesium, silicon, calcium, and iron. The second group was found to be most unusual and is called (inappropriately for chemists) the "CHON" particles, where CHON is an acronym for the presence of carbon (C), hydrogen (H), oxygen (O), and nitrogen (N). The origin of the CHON species was attributed to an unknown process in which simple carbon-containing compounds in the nucleus are formed by ionizing radiation.

A further indication of the existence of specific chemical processes was provided by multicolor photographs taken by Giotto, showing that the surface of the nucleus is "dark gray". Only a very small part of the light that falls on the nucleus is reflected. The degree of reflection is one of the lowest known in the Solar System and is matched only by certain carbon-rich asteroids and the dark lava flows on the Moon. It has been suggested that a tar-like organic material, such as polymers of cyanide, might form a coating on the surface. Alternatively, it was suggested that there might be some kind of spongy texture in which incident light is trapped by multiple internal reflections.

The close encounter with Halley's provided precious information, but it also gave rise to new puzzles. What causes the jet emissions and why is so little of the surface active? What is the origin of the dark-gray material on the surface and of the CHON species in the dust? Are we looking at primitive presolar material or the altered primordial ingredients of the Solar System?

RADIATION AND COMETARY CHEMISTRY

Bathing in Ionizing Radiation

The comet nucleus is exposed to various types and intensities of ionizing radiation throughout its lifetime. This occurs from within through the decay of imbedded radionuclides, and from without by the action of cosmic rays.

We have seen in Chapter 5 that the explosion of a supernova very likely occurred at the beginning of the accretion process of the presolar nebula and the formation of the Solar System. Nucleosynthesis during the explosion produced a large variety of radionuclides, which were incorporated into the comet nuclei in the same way as in other bodies of the Solar System. The radiation emitted during their decay provided a source of energy for chemical processes. Even at present, the surface layers of a comet undergo continuous chemical alteration resulting from the action of electromagnetic radiation and beams of charged particles from stellar wind and cosmic rays.

The intensities of these radiations are fairly weak and the doses delivered are mostly only a few tenths of a gray per year, but the time available for irradiation is very long (thousands of millions of years) and it is the total dose which is important for radiation chemistry. Moreover, the environmental conditions (low temperature and low density of matter) are almost ideal for an accumulation of radiolytic products.

In order to gain insight into the types and extent of radiation-induced changes one must first take into account the total amount of ionizing energy utilized, i.e., the absorbed dose, which is often expressed in gray units (Chapter 3). To calculate the absorbed dose due to the imbedded radioactive nuclides it is sufficient to consider only the main contributors, i.e., the radioactive elements which have survived the process of accretion of presolar nebula. These are ^{10}Be, ^{40}K, ^{129}I, ^{232}Th, ^{235}U and ^{238}U, ^{237}Np, ^{244}Pu and ^{247}Cm. The calculation is based on the amount of radioactive element, its half-life and the energy liberated per radioactive decay. It can be shown that these nine radioactive isotopes furnished a total dose of about 2.80 megagrays in the course of 4600 million years of the comet's life.

Of the numerous shorter-lived species which might also have been present during the formation of bodies in the Solar System, we consider only ^{26}Al, whose half-life is 0.7 million years. Its contribution to the total dose is 10.98 megagrays.

From theoretical considerations, the radiation of ^{26}Al could also have been a source of sufficient radiogenic heat to maintain the core of larger comet nuclei in a liquid state. In this case, the dose mentioned should be particularly significant, since radiation chemical yields are considerably higher in the liquid than in the solid state.

Cosmic radiation consists both of electromagnetic radiations and charged particles, but the calculation of absorbed doses can be simplified by considering only the high-energy protons, which have a predominant role in the energy deposition. It follows that during the comet's life the surface layers receive enormous doses: from 300 megagrays in the first few centimeters to 0.3 megagrays at a depth of 20 meters.

If we also introduce into the calculation the effect of other components such as electrons, electromagnetic rays and the particles heavier than protons, the dose-depth distribution changes significantly only down to 1 meter below the surface. The total dose is doubled in the first 10-centimeter layer, but this contribution decreases rapidly and is only 1 percent of the overall amount at a depth of 2 meters.

The impact of high-energy protons affects a considerable number of atomic nuclei during the collision cascade process in which the proton's energy degrades. The number of affected nuclides decreases rapidly from 10^{14} per gram in the first 10-centimeter layer to 10^{11} nuclides per gram in a 10-centimeter layer which is located at a depth of 5 meters. Each affected nucleus is responsible for radiation damage in the surrounding material; this may involve displacement of atoms and the appearance of vacant lattice sites, or a large local heating effect (hundreds of degrees kelvin) within a small volume (of the order of a cubic micrometer).

Nuclear reactions which take place during the passage of cosmic rays can also form radioactive nuclides. Radiation from this source contributes only slightly to the total dose, amounting to 0.16 megagray in the first 10 centimeters.

Total doses due to cosmic radiation are very high in the outer surface layers. On the other hand, only a small fraction of the total mass of the comet is involved, since the radiation process occurs only in the outer part of 20 to 25 meters of thickness. The only radiation factor which really affects the comet as a whole is the contribution arising from the decay of imbedded radionuclides.

Extent of Chemical Alterations

Radiation-induced chemistry in the comet nucleus essentially involves the radiolysis of water ice in a frozen aqueous system. An estimate of the amount of material altered by radiation at a given dose is made by assuming an average radiation chemical yield (G) and an average molecular weight (M) of compounds produced in this way. One percent of cometary ice material is chemically altered for a dose of 10 megagrays if $G = 0.1$ and $M = 100$. This means that, owing to very large doses, the presolar material in the first meter of surface layer of the comet nucleus is completely reprocessed by radiation after 4600 million years. Similarly, alteration to the extent of a few percent would be expected at a depth of 10 meters, and less than 1 percent below 20 meters.

The existence of liquid water in the nucleus would cause more extensive alterations because of the higher mobility of free radicals and higher radiation chemical yields. Further, it would provide better conditions for the syntheses of compounds with higher molecular weights. Assuming $G = 0.45$ and $M = 1000$, and that only 40 percent of the radiation energy of ^{26}Al was utilized in the liquid core (the rest being dissipated in ice before and during melting), one calculates that about 14 percent of the cometary liquid core could have been altered. The total amount of radiation-processed cometary material in such a case would be very important, since the radius of the liquid core is expected to have been 7 kilometers for a comet of 20 kilometers cross section.

In our example, the choice of values for the radiation chemical yield and molecular weight was based on the somewhat scanty information of simulation experiments, and the above estimates provide only a very general idea of the extent of alterations produced by radiation.

Possible Radiolytic Products

The simulation experiments are performed with accelerated protons or electrons, or with the gamma-rays of ^{60}Co. The target consists of a frozen mixture of several volatile compounds which are assumed to be the main constituents of a comet nucleus. The radiolytic products are identified by various physical and chemical measurements, either *in situ* during irradiation or, more frequently, following thawing of the irradiated sample. The chemical reactions which occur are complex and involve the participation of dozens of free radicals in the formation of as many compounds, some of which are biologically significant.

Experiments have been carried out using a model system containing, in addition to water, the simple organic molecules hydrogen cyanide (HCN), methanol (CH_3OH), methyl cyanide (CH_3CN), ethyl cyanide (C_2H_5CN) and formic acid (HCOOH), in proportions in which these exist in a dense interstellar cloud. Their total amounts were selected in accordance with generally accepted facts on the chemical composition of a comet nucleus. The mixture contained water as a main component and the carbon to nitrogen ratio was 1:8. Irradiations at megagray doses of γ-rays were carried out at 77 degrees kelvin, the temperature of liquid nitrogen, and the samples were analyzed after thawing. Comparative investigations were also performed at room temperature. They provided not only valuable insight into the chemistry of the liquid core, but also into low-temperature processes.

The list of radiolytic products found is impressive, containing about 40 compounds which include molecular hydrogen, hydrogen peroxide, carbon dioxide, carbon monoxide, methane, ammonia, amides, amines, aldehydes, carboxylic acids and larger molecules with molecular masses up to 80,000 atomic mass units. Following hydrolysis by methods used in protein chemistry, it was possible to identify several protein and nonprotein amino acids as well as traces of the biologically significant molecule adenine.

Chemical and Physical Implications

Molecular hydrogen is continuously generated by radiation and can accumulate in all parts of the comet nucleus. It is very likely that it plays a significant role in the outgassings such as are observed in Halley's comet and in certain others even at great distances from the Sun.

The simulation experiments show that the primary constituents become depleted as the dose increases, whereas the radiolytic products formed tend to accumulate. The latter eventually become involved in the chemical process and gradually decompose, giving way to new radiolytic products. This suggests that there must be a nonhomogeneous formation and accumulation of radiolytic products within the icy core of the comet. The expected variation of dose with depth therefore points not only to variations in the amounts of energy absorbed and compounds synthesized, but also to differently advanced stages of radiolysis and different types of compounds subsequently accumulated. These respond differently to the effect of heat when the comet approaches the Sun, and may be responsible for the anisotropic emission of gas and dust as has been observed in Halley's comet.

Experiments also show that the ionizing radiation leads to formation of a dark-colored mixture of nonvolatile compounds containing several polymers. This may correspond to the "friable sponge" material that is thought to cover the crusty surface of Halley's comet and is considered responsible for its low albedo.

On the whole, the simulation experiments suggest that the primordial material of the comet is a remnant of a presolar nebula which has been chemically altered to various degrees by cosmic rays and radiations from radionuclides imbedded in the cometary material.

Evolutionary Implications

Comparative planetology shows that the primitive Earth must have been submitted to heavy bombardment by meteorites and comets. This suggests that cometary impacts may have deposited organic compounds, and thus the possible role of comets in prebiotic evolution must be taken into consideration. It has been estimated that if only 1 percent of cometary organic material survived the impacts, the mass of carbon-containing substances which reached the Earth should have been one or two orders of magnitude greater than the total organic mass in the presently existing biota. Similarly, if only 10 percent of the impacting mass had been carried by comets, the amount of water contained would have been sufficient to form oceans.

A more provocative idea was put forward by the English astronomer Fred Hoyle: the first form of life originated in cometary nuclei and could have reached our planet along with them or their debris. He also imagined a widespread presence of bacteria and viruses in the interstellar medium. These conceivably caused worldwide epidemics in the past when the Earth's population was scanty and human contacts were rare. Such ideas appeal to the general public and are readily accepted by the mass media; the scientific community tends to remain sceptical. It is difficult to see how life as we know it, based on the chemistry of carbon and liquid water, could have emerged and survived in interstellar space, where the temperature and density of matter are very low and liquid water is virtually absent. Further, how could living forms have survived for millions of years in the hostility of cold storage and energetic radiation, before reaching the aqueous environment of Earth for further evolution.

METEORITES, VISITORS FROM THE INNER SOLAR SYSTEM

Lumps of Rock and Metal

All rocks that fall from the sky onto the Earth or other bodies in the Solar System are called meteorites. Aptly described as the poor man's space probes, meteorites bring us extraterrestrial materials free of charge. They represent relics of cosmic events that have occurred during the past 5000 million years.

Meteorites belong to a larger family of solid celestial bodies called meteoroids that enter the Earth's atmosphere from interplanetary space. They are lumps of rock and metal with sizes ranging from that of a dust grain to the dimensions of a boulder 10 meters thick. The latter kind are rare, and may fall once in several thousand years. Meteorites enter the atmosphere at speeds of 15 to 72 kilometers per second. Their arrival is announced about 100 kilometers above the Earth by luminosity resulting from frictional heating with the atmosphere. Most of them burn up completely and are called meteors or "shooting stars".

It is generally accepted that most meteorites, though not all, come from the belt of asteroids. Some may have been ejected from Mars. Asteroids are much larger lumps composed of rock and metal fragments that revolve around the Sun, mostly in a broad orbit between Mars and Jupiter. At present, about 2000 asteroids are catalogued and named. Their overall mass is smaller than that of the Moon and it is very likely that for the most part, their material was not incorporated into the planets when the Solar System was formed. Occasionally an asteroid crosses the Earth's orbit, as in 1972 over the U.S. and Canada. This object had a diameter of about 10 meters, a weight of 1000 tons, and a speed of 15 kilometers per second. It was visible to the naked eye by daylight. The escape of an asteroid from its belt results from gradual changes in its orbit.

A more common event is the appearance of spectacular meteor showers, which are probably of cometary origin. The remnants of bygone comets travel in swarms around the Sun and appear year after year in the night sky. They may be reminiscent of a shower of snowflakes in a driving head wind, and thousands can be seen in the course of an hour.

Micrometeorites, which float gently to the ground, represent another category of falling objects. They are probably composed of cometary dust particles which are captured by the Earth's magnetic field.

The total mass of meteoritic matter that settles daily on the Earth is estimated at about 100 tons, but only very little of it is seen since most of it consists of micrometeorites. Moreover, many objects fall into the oceans or on remote and uninhabited areas. Indeed, some of the more common types of meteorites cannot be distinguished from terrestrial rocks.

The existence of craters appears to be generally widespread in the Solar System and, as comparative planetology suggests, the impact of much heavier and larger meteorites was a common and frequent occurrence in the distant past. About 80 places are known on Earth where larger meteorites have fallen. Of particular interest is a crater in Arizona (U.S.), which is almost 1.5 kilometers wide and which indicates that about 400 million tons of rock was pulverized by the impact.

The largest known meteorite to date is located in Namibia (Africa). It weighs 60 tons and is still kept on the spot where it was found. The biggest one exhibited in a museum is the Ahnighito, a hunk weighing 34 tons that constitutes a centerpiece at the American Museum of Natural History.

Obviously, many tons of meteoritic material are available: about 2000 specimens are carefully preserved in museums and scientific collections around the world. For their analyses, highly specific and sensitive methods are used, in particular, techniques based on radioactivity for determinating their age, and the duration of exposure to cosmic rays.

Composition

Whatever their origin, meteorites are classified according to their composition into three main groups: iron, stony, and stony-iron.

There is no terrestrial material on the surface of our planet that resembles an "iron" meteorite, which consists of an alloy of iron with 4 to 16 percent nickel. The only place where anything like this might exist is about 5000 kilometers beneath our feet, in the semimolten core layers of the Earth. Stony-iron meteorites, like their iron counterpart, are also different from any rock found on Earth.

The stony meteorites represent by far the most abundant class and are divided into two major groups. The chondrites (85 percent) are unique in containing chondrules, tiny blobs of glass and crystals about 1 millimeter in size. The achondrites (15 percent) are very much like igneous earth rocks called brecias. Numerous samples of brecias similar in composition to achondrites were brought from the Moon.

Of particular interest for studies of chemistry and evolutionary processes in interplanetary space are the carbonaceous chondrites, a smaller group representing several percent of the meteoritic stock. They are considered to be the most primitive of their kind. They may contain up to 5 percent carbon and 20 percent water; molecules of the latter are structurally bound in the mineral context. These mineral inclusions are believed to be the oldest in existence, probably having been formed when matter in the presolar nebula started to collapse and constitute the material from which all planets and the Sun evolved.

Most of the organic matter in carbonaceous meteorites (70 to 95 percent) is present as an ill-defined, insoluble, macromolecular material. In some respects it resembles organic polymers like humic acid in the soil. Electron diffraction measurements suggest that up to 80 percent of the carbon in some carbonaceous meteorites may be present in the form of carbyne. This is a polymorphic compound of elemental carbon based on a chain structure with alternating single and triple bonds, and represented as $-C{\equiv}C-C{\equiv}C-C{\equiv}C-$.

Fairly complex organic compounds were also found in the carbonaceous chondrites, and these meteorites are considered as a potential source of biologically important molecules on the primitive Earth. Analyses show the presence of about 40 protein and nonprotein amino acids after hydrolysis. The most abundant is glycine; amounts of other amino acids decrease as the number of carbon atoms in their molecules increases. Appreciable amounts of monocarboxylic acids and hydrocarbons are also found, as well as traces of purines and pyrimidines.

Analyses strongly suggest that all these compounds are indigenous. Some of them may already have been present in interstellar dust, but the abiotic, chemical synthesis during and after accretion of presolar material was presumably the main source. Because of penetrating ionizing radiation and the chemically effective free radicals which they formed, the role of radiation processing might have been particularly significant.

Exposure Age

In the same way as comets and other unshielded bodies in interplanetary space, meteorites are permanently exposed to bombardment by cosmic rays. Nuclear reactions take place along the pathways of cosmic rays down to 1 meter or more from the meteorite's surface. The number of species formed there, both stable and radioactive, can be used in estimating the time of exposure of the material to cosmic rays. In principle, the larger the number of radioactive species, the longer the exposure time was. In practice, however, the matter is more complicated. For exposure times longer than the half-life of the measured radionuclide the induced radioactivity assumes a limiting value when the decay rate becomes equal to that of formation. In this case, supplementary information is required, and this is provided by the concentration of the stable product of radioactive decay that continuously accumulates. Pairs of nuclides which are of interest in this respect are ^{36}Cl (half-life 300,000 years) and ^{36}Ar (stable), or ^{22}Na (half-life 2.6 years) and ^{22}Ne (stable). Using these pairs, it is possible to calculate the exposure age at the time the meteorite fell. In the so-called meteorite "finds", i.e., meteorites whose fall was not observed, the apparent disagreement between the results obtained with different pairs can also be used to deduce the time which elapsed since the meteorite's fall.

Numerous exposure ages of meteorites have been determined and have provided an insight into various events in the interplanetary medium. The exposure ages of some stone meteorites range from 20,000 to 80 million years, while for iron meteorites they vary from 4 million to 2300 million years. The generally much shorter exposure age of stone meteorites may mean that they originated from parent bodies different from those of iron meteorites. Alternatively, stone meteorites could have been broken up in subsequent collisions, which is conceivable because of their greater fragility. A pronounced clustering of exposure ages appears for some classes of meteorites. This may be supporting evidence for their origin through breakup of a common, larger parent body.

Age of Meteorites

The most frequently used radioactive clocks for dating meteorites are the decay processes ^{40}K to ^{40}Ar, ^{87}Rb to ^{87}Sr, and ^{147}Sm

to ^{143}Nd. The oldest meteorites which have been reliably dated are about 4600 million years old, which is about 100 million years older than the oldest lunar rocks, and also somewhat older than our planet.

The determination of the age of meteorites using a radioactive clock will be illustrated for chondrites. The clock is based on the beta decay of ^{87}Rb to ^{87}Sr with a half-life of 4.88×10^{10} years. When chondrites started to crystallize during the accretion of the solar nebula, the minerals incorporated rubidium and strontium. Owing to the decay of ^{87}Rb, the amount of ^{87}Sr increases steadily as time passes. However, the age cannot be inferred from the simple determination of ^{87}Sr, since at the time of its formation the meteorite already contained this isotope, as well as other strontium isotopes such as ^{86}Sr. Accordingly, it is the increase with time of the ratio ^{87}Sr to ^{86}Sr which must be taken into account, whereby the initial value of this ratio is 0.6968.

In a typical procedure, several separate fractions from each meteorite are analyzed for ^{87}Sr, ^{86}Sr, and ^{87}Rb. Each fraction contains different minerals or at least different amounts of the same mineral. A graph of the ^{87}Sr/^{86}Sr ratio vs. the ^{87}Rb/^{86}Sr ratio for each specimen yields a straight line called an isochron, because all samples lying on the line have the same age, which is derived from the slope (Figure 24).

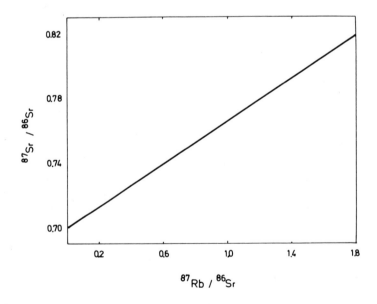

FIGURE 24. Age determination of chondrites with the ^{87}Rb/^{87}Sr radioactive clock.

The isochron shown in Figure 24 was obtained from data based on a large number of chondrites and indicates 4498 million years (±15 million) as the "best" value for the age of meteorites. It can also be deduced that all chondrites must have been formed simultaneously from the solar nebula and that the primordial strontium isotopes were distributed homogeneously, since the initial ratio $^{87}Sr/^{86}Sr$ is constant.

The crystallization age is another feature of interest. Radioactive dating clearly shows that the crystallization age is about 4600 million years, i.e., as old as meteorite formation at the commencement of the Solar System.

Dating Events in the Solar System

Radioactive dating of meteorites also provides evidence that some of the chunks that fell from the sky originated in more recent and quite different events. Radioactive clock determinations based on the samarium-neodymium and rubidium-strontium pairs clearly suggest that in some cases the crystallization age is only a few hundred million years. From additional physical examination it appears that the material was part of a lava flow at the time of crystallization. These observations led scientists to propose the following event: in the relatively "recent" past, about 200 million years ago, a huge meteorite struck solid lava somewhere in the Solar System and projected chunks into interplanetary space. This interpretation is rationalized by the fact that the material bears signs of quite recent severe shock waves.

Where this event occurred, and what the parent body was, remains a matter of speculation. Lunar exploration shows that the Moon can be ruled out. Our satellite has been a geologically dead body for quite a while, the last lava flux having occurred some 2500 million years ago. By successive elimination Mars is left as the most plausible site. Observations of its surface show that lava flows occurred there in the relatively recent past, about 200 million years ago. A possible way in which material could be ejected into space would be by powerful meteoritic impact resulting in instantaneous evaporation of a large amount of frozen water in the Martian soil, followed by explosive steam propulsion of fragments into space. We may have been handling the chunks of Martian rocks for years without realizing it!

Further Reading

Delsemme, A.H., Ed., *Comets, Asteroids, Meteorites,* University of Toledo, Ohio, 1977. [Scientific papers from an international meeting where interrelations, evolution, and origins of comets, asteroids, and meteorites were presented. A useful survey of the state of the art up to the mid 1970s.]

Dullet, W.W., and Williams, D.A., *Interstellar Chemistry,* Academic Press, New York, 1984. [An introduction to the foundations prepared for physicists and chemists who intend to undertake research in this area. Offers a general and up-to-date view and can be used by nonprofessionals who have a more profound interest in this new branch of chemistry.]

Gallant, R.A., *Our Universe,* National Geographic Society, Washington, D.C., 1980. [A concise, clearly written text with beautiful illustrations. A book that should be in every library.]

Nagy, B., *Carbonaceous Meteorites,* Elsevier, Amsterdam, 1975. [Rich in details. A sourcebook on the most primitive rocks which fall from the sky. Of interest not only to scientists studying meteorites.]

Ponnamperuma, C., Ed., *Comets and the Origin of Life,* D. Reidel, Dodrecht, 1981. [Scientific papers from an international meeting at which comets were examined from various aspects in particular their contribution furnishing volatile and organic compounds to the early earth and to processes leading to the origin of life. The texts are prepared by scientists for scientists, but can be understood by laymen having basic knowledge in chemistry and astrophysical sciences.]

Wood, J. A. and Chang, S., Ed., *The Cosmic History of the Biogenic Elements and Compounds,* NASA Scientific and Technical Information Branch Washington, D.C., 1985. [A report by a study group of experts clearly written and illustrated; prepared as a scientific guide.]

MacSween, H.Y., Jr., *Meteorites and Their Parent Planets,* Cambridge University Press, Cambridge, Massachusetts, 1987. [A clearly written and well-illustrated presentation of meteorites and their parent bodies].

Whipple, F.L., *The Mystery of Comets,* Cambridge University Press Cambridge, Massachusetts, 1985. [A popular presentation of progress in the field of cometary science from ancient times to the present era. Written by one of the pioneers in the subject, the book is full of fine personal recollections.]

A popular presentation on meteorites with excellent illustrations has appeared in the special supplement of *Natural History,* Vol. 90, April 1981.

A survey of scientific results obtained during the close encounter with Comet Halley in 1986 appeared in *Nature,* Vol. 321, May 15, 1986, and in several publications of the European Space Agency (ESA, 1986 and 1987).

Our planet was born about 4500 million years ago, after the accretion of presolar nebula ended and the Solar System had appeared. From what we know or assume about the presolar nebula, accretion took place very likely in the presence of abundant radiation and radionuclides. The radionuclides were incorporated into the planet's body and their decay makes possible not only the dating of various events in its past, but also a reconstruction of some cosmological events that preceded the birth of the planet. The energy liberated during radioactive decay was an important source of terrestrial heat and an efficient means of promoting evolutionary processes.

The ionizing radiation arose not only from radionuclides dispersed in terrestrial material, but also from localized uranium ore deposits which possessed certain similarities with present-day man-made nuclear reactors. One such fossil reactor site which had remained in obscurity for 2000 million years was discovered in the 1970s during the mining of a uranium deposit at Oklo (Gabon, West Africa). Chemical analyses of thousands of samples taken from the reactor site, together with computer modeling of experimental findings based on the nuclear reactor theory, have provided insight into the working conditions of the six fossil reactor cores. The phenomenon of Oklo is surely not unique in the past history of our planet. From available information it is possible to reconstitute the picture of a natural nuclear reactor on the early Earth. The number of such reactors may be roughly estimated from the amount of uranium in the Earth's crust that was possibly involved in the criticality.

Chemical evolution was a sequence of events on the early Earth which led from simple inorganic molecules to complex organic compounds and eventually to primitive forms of life. It was followed by biochemical (early biological) evolution. Ionizing radiation was one of several types of energy that were available for these evolutionary processes. Radiation-generated free radicals in bodies of water covering the primitive Earth were efficient in promoting important chemical reactions for both abiotic syntheses during chemical evolution and biochemical processes in the primitive forms of life during early biological evolution.

7

The Early Earth

FAR INTO THE PAST

Radioactive dating of meteorites and lunar rocks shows that the birth of these bodies in the Solar System occurred about 4600 million years ago. It also indicates the age of processes leading to the formation of our planet Earth.

Materials as old as this are unknown on the Earth despite numerous datings of rocks and minerals and the reliable techniques available for geological dating. Of particular help is the use of certain primordial radionuclides as radioactive clocks, in which the age is derived from measurements of radioactive decay of long-lived radioisotopes such as ^{87}Rb, ^{147}Sm, ^{238}U, or ^{235}U. In only a few dozen locations (mainly in Africa, Australia, and North America), the measured ages correspond to origins which date as far back as 3000 to 3500 million years ago.

However, some individual minerals, as distinct from rock formations, have reliably established ages of the order of 4300 million years. These were found in Australian sandstones, sedimentary rocks which consist chiefly of cemented quartz grains (quartzite), and which were deposited and methamorphosed about 3800 million years ago. Their age was derived from the content of uranium, thorium, and radiogenic lead.

183

The oldest rock on Earth has an age of 3800 million years (Isua, Greenland). It is not unlikely that some older rocks might be found in the future, but they will not be much older than the Isua rocks and will not be widespread. The main reason for the lack of primordial remnants is that traces of the Earth's primitive crust have disappeared in dramatic events which shaped its surface during the early stages in the constitution of the Solar System.

Comparative planetology, a new branch of science which has developed with recent exploration of the Solar System, offers insight into the latter's past. This discipline suggests that formation of the protocrust was followed by intense tectonic and volcanic activity and that showers of comets and meteorites must have heavily bombarded the Earth's surface. Impact craters on Mercury are an example of well-preserved remnants of activity in the early Solar System. Studies of lunar rocks show that the intense bombardment must have ceased about 3900 million years ago. Heavy rains and winds contributed towards an effective erosion of the formerly lifeless surface of the planet. Traces of important erosion on the surface of Mars, and observations of complex chemistry in the atmosphere of Venus, are of great assistance in the attempt to reconstruct the earliest stages of geophysical and geochemical evolution of the Earth.

The Oldest Known Rocks

The Isua Belt of West Greenland is famous because of the rocks which are about 3800 million years old. Since the early 1970s when their first dating was reported, these rocks have been considered to be the oldest terrestrial materials to have survived under relatively pristine conditions until the present time.

The dating of this sediment was accomplished with the aid of several parent-daughter pairs, and measurements of isotopic ratios of the decay products have provided values for the rock's age which are in very good agreement: 3770 million years (^{147}Sm/^{143}Nd); 3760 million years (lead isotopes ratio), and 3750 million years (^{87}Rb/^{87}Sr).

The Planet's Birth

In order to interpret various findings with respect to extinct nuclei and their radioactive parents, i.e., isotopic anomalies, it has been proposed that a supernova explosion may have occurred some time before or near the beginning of the accretion process of the presolar nebula.

Isotopic anomalies could also have other causes. Arguments against the role of a supernova explosion have been advanced and the hypothesis is far from being generally accepted. However, the arguments are not sufficiently convincing to rule out the supernova hypothesis when the potential radioactivity on the early Earth is taken into consideration. The reason for this is that a supernova explosion could trigger the synthesis of many chemical elements, including those of high mass number. A survey of the properties of all known nuclides, including those in nature and the ones produced in the laboratory, shows that almost 80 percent are radioactive; consequently, about as many would also be expectedly radioactive under conditions of nucleosynthesis. Their half-lives vary within a wide range, from fractions of a second to 1000 million years and more. During accretion of presolar material the radionuclides were very efficiently incorporated into the planetary bodies; nuclides with half-lives longer than 1 million years are of particular interest in the terrestrial context. They represent a significant 1 percent of the total number of synthesized nuclides. On the present-day Earth we still find about 15 long-lived radionuclides which are the remnants or descendants of radioactive nuclei synthesized in supernova explosions in cosmic space. Some may have originated in such an explosion associated with the accretion of the presolar nebula.

Further information which also supports the supernova hypothesis is provided by the ^{129}Xe anomaly in meteorites. The isotopic analyses of meteoritic material reveal an excess of ^{129}Xe which is attributed to the decay of the extinct ^{129}I (half-life: 17 million years). Consequently, this radioiodine must have been present during the accretion of the cosmic dust cloud from which the Solar System was formed. It also implies that the ^{129}I incorporation into the solar bodies was of a relatively short duration. More precisely, it may be inferred that less than 160 million years must have elapsed; otherwise ^{129}I would have decayed and the xenon isotopic anomaly would not exist.

Some important events in the past of our planet are surveyed on Figure 25.

ATMOSPHERE AND CRUST

A continuous production of cosmogenic radionuclides occurred in the primitive atmosphere. Radioactive nuclides were also dispersed throughout the primitive planet. They included many of those which we presently find in laboratory experiments or in nature, and, possibly, the superheavy elements. The total extent of radioactivity was much larger on the primitive Earth, since there has been significant decay of radioactivity during the 4500 million years of the planet's life.

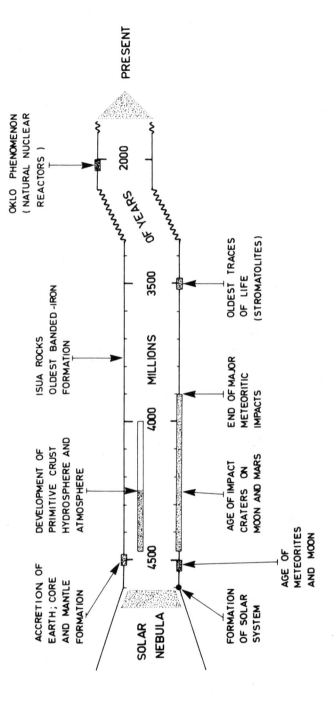

FIGURE 25.   Some events relating to the origin and early evolution of the Earth.

Cosmogenic Radionuclides

The actual composition of the early atmosphere can only be surmised, since the oldest known rocks are not sufficiently old to tell us about the chemistry at the surface of the primitive planet. However, the atmospheres of Jupiter and Saturn are well known. They most likely still correspond to the primitive atmospheres, which remained in a state of conservation because of both the large masses and gravitational forces of these celestial bodies and their considerable distances from the Sun. It is probable that the primitive atmosphere of the Earth consisted mainly of hydrogen, both in molecular (H_2) and in chemically combined form with the most abundant cosmic chemical elements: carbon (methane, CH_4), oxygen (water, H_2O) and nitrogen (ammonia, NH_3). This atmosphere may have already existed during the final stages of condensation of the presolar nebula, and perhaps also for some time after the core, the mantle, and the crust were formed. Later, owing to the smaller mass and weaker gravitational force of the Earth, its primitive atmosphere was lost. It seems that the Earth rapidly replaced its primary atmosphere by a process of repeated degassing, which occurred during the terminal stage of accretion and thereafter during intense impacts and volcanic eruptions. In addition to water vapor and slight amount (1 percent) of hydrogen (H_2), this new atmosphere also contained nitrogen (N_2), carbon monoxide (CO), and hydrogen sulfide (H_2S). For reactions leading to the formation of cosmogenic nuclides the same main target nuclei were present as in the atmosphere today, namely, hydrogen, oxygen, carbon, and nitrogen. Thus, the same radioactive nuclides are characteristic for both the primitive and present atmospheres of the Earth.

Their respective amounts, however, must have been different because of different prevailing intensities of cosmic rays. The cosmic ray flux was probably more intense in the early stages of the Solar System. One reason is that 4600 million years ago, the frequency of supernova explosions in the galaxy may have been higher than that observed at the present time. Also, as present observations of young stars show, the intensity of all radiations (both electromagnetic corpuscular) seems to be much higher during the first stages of the star's life.

Radiogenic Heat

When the radiations emitted during radioactive decay are absorbed in matter their energy is eventually converted into heat. This radiogenic heat was, and still is, of importance for the heat balance of the Earth.

During accretion of the presolar nebula and the formation of the Earth, the heaviest radioelements were drawn toward the center of the planet. The energy released from their radioactive decay, together with gravitational energy, was the source of heat that led to the formation of the molten core during the first 100 million years of our planet's existence.

Uranium, thorium, and ^{40}K are of particular importance as sources of radiogenic heat (Table 13). These elements are not confined to the planet's interior and are found everywhere in the crust. It seems that they migrated to the Earth's surface at a very early stage as a result of their crystalline and chemical properties. Various laboratory studies suggest that they were highly soluble in the molten material that moved from the liquid core toward the surface.

Table 13

RELATIVE ABUNDANCES OF PRIMORDIAL RADIOACTIVE ISOTOPES IN THE ELEMENTS URANIUM, RUBIDIUM, AND POTASSIUM

		Relative isotopic abundance (%)	
Nuclide	**Half-life (years)**	**Early Earth (4500 million years ago)**	**Present Earth**
Uranium-238	4.47×10^9	74.020	99.274
Uranium-235	7.04×10^8	25.990	0.720
Rubidium-87	4.88×10^{10}	29.19	27.83
Potassium-40	1.28×10^9	0.145	0.0117

Note: For uranium, which has no stable isotopes, the data represent the isotopic composition of the element.

Actually, if such migration had not occurred, the geological behavior of our planet would have been quite different. The confinement of radioelements within the core would have led to the generation of more heat and to an increase in the temperatures of the core and mantle. In consequence, tectonic activity in the crust and earthquake frequency would have considerably increased.

Other radioactive nuclides, with low abundances and very long half-lives, also contribute to the generation of heat despite a much weaker degree of radioactivity. One of these is ^{87}Rb, which represents 28 percent of rubidium in nature; this element is fairly common in the Earth's crust (78 grams per ton). ^{87}Rb emits weak beta (β)-rays and decays with a half-life of 4.88×10^{10} years.

Another chemical element of interest is indium (0.2 gram per ton). About 96 percent of this element consists of ^{115}In, which emits β-rays and has a half-life of 5×10^{14} years. Samarium exists to the extent of 7 grams per ton of the crust: about 26 percent of natural samarium is a mixture of three radioactive isotopes with half-lives ranging from 10^8 to 10^{16} years.

The total energy attributed to radioactive elements is, however, fairly low. Calculations based on the radioactivity content of a 1-kilometer layer of the crust show that 1 square meter of the surface of the primitive Earth received 117 kilojoules per year. The corresponding value is 33.4 kilojoules for the present-day Earth.

Superheavy Elements

Theoretical considerations indicate that a group of superheavy elements should exist with the atomic numbers in the region of 114 (Table 1, Chapter 1). These would be expected to decay by spontaneous fission, or by α- and β-decay and have half-lives ranging anywhere between seconds and minutes to 1000 million years.

The most stable and tightly bound nuclei are those in which the protons and neutrons fill up a nuclear shell in a manner similar to that of electrons filling atomic shells. By extrapolating from the known data for existing nuclei, theorists predict that nuclei with about 114 protons and 184 neutrons might have much longer lifetimes than those of the neighboring nuclides. These superheavy nuclei may have been formed during nucleosynthesis, and if a supernova explosion indeed occurred at an early stage of accretion in the presolar nebula, one would expect them to be present among radioactive nuclides. Hence, traces of some of the longest-lived elements might still be present on the planet. By the end of the 1980s the search had been continuing for almost 2 decades without furnishing a positive answer as to their existence in nature. At the same time, artificial production of transuranium nuclei has not progressed beyond atomic number 109.

THE PHENOMENON OF OKLO

The first artificial nuclear reactor was put into operation in 1942 following a complex program of research involving the coordinated efforts of almost 100,000 scientists and technicians. For 30 years it was believed to be the first nuclear reactor on our planet, until, in 1972, a team of French scientists discovered a "fossil" nuclear reactor site at a uranium mine in Oklo (Gabon, Africa). This discovery revealed that, already 2000 million years ago, at least six nuclear reactors were in operation. These were natural arrangements of uranium and water in which, as in the present man-made nuclear reactors, the fission chain reaction of uranium took place. Natural conditions of criticality were easier to attain in the early history of our planet owing to a higher abundance of fissile uranium in the nature (up to 26 percent); at present it is only 0.72 percent.

A search was made for uranium chain fission processes in nature even before the discovery at the Oklo uranium mines. Investigations have shown that, when extracted from an old and well-preserved uranium ore deposit, uranium minerals indeed contain traces of fission products. However, there was no evidence that the latter were formed in chain fission reactions. In fact, they are primarily attributed to the products of spontaneous fission of ^{238}U and, to a lesser degree, to the fission of ^{235}U induced by neutrons released in the spontaneous fission of the heavier uranium isotope.

Discovery

The discovery of the fossil nuclear reactor site at Oklo was not a consequence of the search for uranium chain fission processes in nature and an account of the events somewhat resembles a good detective story.

The first evidence came from an extraordinary combination of chance, extreme technical care and scientific ingenuity. A routine control of natural uranium, which was to be treated in a French factory for the enrichment of ^{235}U, had revealed a small anomaly in its content of fissile uranium: the measured value was 0.7171 percent instead of 0.7202 percent, which is accepted as the standard value for the atomic concentration of ^{235}U in natural uranium. The difference of 0.0031 percent was indeed small, but it was beyond the limits of all possible experimental errors. It was also larger than the discrepancy of 0.0006 percent which was tolerated in French laboratories. In consequence, the anomaly was taken seriously and subjected to further examination.

Scientists knew that the concentration of fissile uranium in samples taken from various locations on the globe is constant up to a few parts per thousand, and the observed minor discrepancy was tentatively attributed to some uncontrolled mixing of natural uranium with uranium that had already been used in a nuclear reactor and, therefore, was impoverished in the fissile isotope.

The first task was to determine at what stage dilution with depleted uranium might have occurred. The anomaly was observed at one of the final stages of uranium processing, in the enrichment. Clearly, the search had to go backward through time and the sequence of procedures, that is, through the complex succession of operations in which several organizations were involved. To the surprise of the investigators, the trace led to certain ores from the Oklo uranium mines in Gabon, a small country in the western part of equatorial Africa. The detailed analyses of samples taken from the uranium-rich zone in the ore body were even more surprising: the samples which had the highest content of uranium were also those in which the fissile isotope was most depleted. In some cases the concentration of ^{235}U was only half of the standard 0.72 percent, or even less. Further analyses also revealed the abundant presence of the rare earth elements. The latter are known to appear as the end products of radioactive decay of fission-formed radionuclides. All these findings suggested that a high degree of nuclear combustion must have occurred under natural conditions. The conclusion was reported to the French Academy of Sciences and published in its *Comptes Rendus* in 1972.

But that was not the end of the story. Complex multidisciplinary investigations were carried out in the years following the discovery. Two international scientific meetings on the Oklo phenomenon and on natural nuclear reactors were organized by the International Atomic Energy Agency. The proceedings of these meetings offered a detailed picture of the geology and geochemistry of the fossil reactor site. They provided insight into the pathways of the formation of uranium deposits where criticality had been achieved, and proposed an interpretation of the manner in which the uranium deposits with reactor cores were preserved up until the present day. The isotopic analyses furnished detailed information on the ^{235}U consumption and enabled calculation of the energy liberated during the active period of the Oklo phenomenon. Moreover, radioactive dating gave a reliable estimate of the age of the ore deposits and of the time when the fission process occurred.

Chain Fission in Uranium Ore Deposits

We have seen how the uranium nucleus splits into two parts with the release of neutrons (Chapter 4). If the neutrons are sufficiently slowed down (moderated), they can induce the fission of other ^{235}U nuclei. According to the nuclear reactor theory, the conditions for maintaining a chain fission reaction are that, on the average, at least one fission neutron induces another fission. In order to attain the conditions required for triggering and maintaining the chain fission process, a sufficient amount of uranium (a critical mass) must be available and imbedded in a suitable neutron moderator.

As early as 1956, P.K. Kuroda, a Japanese scientist working in the United States, published a short note on the possibility of criticality being reached in natural uranium deposits. He pointed out the higher content of ^{235}U in the past history of the Earth and the suitability of water as a moderator under natural conditions. His calculations suggested that the 2000 million year-old uranium deposits might be the promising places to search for relics of natural chain fission. These ideas were revived many years later, following the discovery of the Oklo site and its fossil reactors.

It is not surprising that the early idea about natural nuclear reactors scarcely attracted attention. The Oklo phenomenon showed that several rarely occurring conditions have to be met simultaneously for a chain process to occur in nature: a rich uranium deposit, the presence of water, and a limited amount of nuclear impurities, i.e., nuclides which are strong neutron absorbers and which interfere with the fission chain reaction. The age of the uranium deposit is another important factor: it must be sufficiently old to contain an appropriate amount of fissile isotope, but young enough to offer convenient geochemical conditions for chemical oxidation and reduction reactions which facilitate the accumulation of uranium. Something else was also needed, which was hardly imaginable before the discovery at Oklo: the site with fossil reactors had to survive drastic changes on the surface of our young planet and remain preserved until our present day.

The Oklo phenomenon proves that it was indeed possible for all these conditions to be met on the early Earth. A critical examination of geochemical and geophysical conditions in the Precambrian period, the eon of the Earth's history from (4.5 to 0.57) $\times 10^9$ years ago, shows that the likely time-span for the occurrence of natural nuclear reactors is rather large: from 4.1 $\times 10^9$ years ago (formation of the crust) up to 1.1 $\times 10^9$ years ago.

Computer simulations of processes in a natural nuclear reactor show that criticality could have been achieved between 4100 and 3100 million years ago in the deposits containing as little as 3.5 to 6 percent of uranium in the ore. This was made possible by the relatively high abundance of fissile isotope, between 20 and 10 percent of ^{235}U in natural uranium. Moreover, this large amount of fissile nuclides could have supplied a sufficient flux of neutrons to trigger and maintain a chain fission process even in the presence of significant amounts of neutron absorbers.

In the more recent past, conditions for criticality required richer uranium deposits in order to compensate for the lower content of fissile isotope. Between 3000 million and 1800 million years ago a natural nuclear reactor needed a geological formation with, respectively, 10 and 30 percent uranium (as a minimum) in the ore. Up to 50 percent uranium content and practically complete absence of nuclear impurities would be needed for chain fission to occur in the recent geological times, i.e., some 1100 million years ago.

Fossil Nuclear Reactors in Oklo

Six reactor cores, spread over a zone 150 meters in length, were discovered at Oklo (Figure 26). The lens-shaped reactors are about 1 meter thick and 10 to 20 meters long. The concentration of uranium in the reaction zones is 20 percent and more, in some cases up to 60 percent; in the surrounding ore matrix it is much lower, being between 0.2 and 0.5 percent. The uranium deposit is a sedimentary sandstone layer whose thickness varies between 4 and 10 meters. It is 2050 million years old, and the chain fission occurred 2000 million years ago.

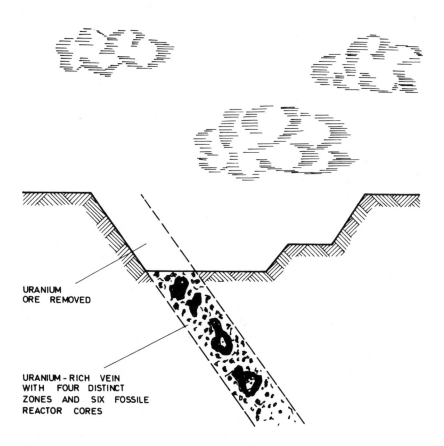

URANIUM
ORE REMOVED

URANIUM - RICH VEIN
WITH FOUR DISTINCT
ZONES AND SIX FOSSILE
REACTOR CORES

FIGURE 26. The Oklo reactor site with six fossil reactor cores in four distinct reaction zones.

An interesting conclusion of the examination of criticality conditions is that the Oklo phenomenon is a result of combined nuclear and geological activities; the geochemical conditions enabled the triggering of the chain reaction and, in return, their occurrence contributed to a further shaping of the geological formation of the reactor site. The sediment with uranium-rich lenses was embedded at great depths during the criticality period, most likely at several thousand meters below sea level. This depth contributed toward preservation of the site following the criticality period. Regression of the sea during the more recent past enabled appearance of the site on the Earth's surface.

For the Oklo phenomenon, the presence of water was of great importance in providing both an efficient moderator of neutrons and a coolant. Because of the high pressures at great subterranean depths, the water remained liquid even above its boiling temperature. It seems very likely that the depth corresponded to about 5000 meters, at which the pressure would have exceeded 200 atmospheres and the temperature of the water could have reached 374 degrees celsius.

The water temperature controlled the chain fission process by modifying the density of the aqueous phase and thus effectively regulating the slowing down of neutrons. As the chain fission process progressed, the released energy accumulated in the water and increased its temperature. If the reactors were embedded beneath a shallow body of water the pressure may have been only about 1 atmosphere. The accumulation of heat would have caused the water to boil and evaporate as soon as the boiling point was reached (100 degrees celsius). At great depths and pressures as in Oklo on the other hand, the temperature would have risen to 374 degrees celsius before the commencement of boiling and evaporation. At that point the water density decreased, the slowing down of neutrons became less efficient, and the rate of the fission reaction also slowed down. In consequence, the water temperature then decreased, leading to a rise in density and a more efficient degree of neutron moderation and eventually the uranium fission was reinitiated. As the power increased, the water temperature rose until shutdown again took place. This cycle of operation and shutdown extended over a period of 600,000 to 800,000 years in the case of individual reactors, and the total duration of all six reactors is estimated at several million years.

The question is still unsolved as to how the first chain reaction was triggered. A possible answer is that at Oklo criticality was propagated like fire in a desert bush. It could have started in one of the reactor cores where the uranium content was highest and (or) the amount of nuclear impurities lowest. For triggering, only a few neutrons produced by nuclear reactions of cosmic rays would have been sufficient. Neutrons escaping from the first reactor zone could have reached a neighboring uranium-rich lens, gradually consuming the neutron poisons in the ore contained and thus facilitating the triggering of a new chain fission reaction. It seems that several linked reactors were involved in the initiation and propagation of criticality at the Oklo site.

The total amount of energy which was liberated during the Oklo phenomenon can be reliably evaluated. The estimate is based on data available for the consumption of fissile ^{235}U. The result is indeed very impressive: 500 million gigajoules (5×10^{17} joules). For comparison, a nuclear power plant with a nominal power of one gigawatt of electricity produces annually about 32 million gigajoules (3.2×10^{16} joules). In the Oklo reactors, however, the energy was gradually produced over a period of a million years of operation. This gives quite a modest value for the annual output. In fact, the average power of individual natural nuclear reactors at the Oklo site was only 2.6 kilowatts, compared with 1 million kilowatts as produced in a contemporary nuclear power plant.

NUCLEAR ENERGY IN THE EARTH'S PAST

The abundance of fissile isotope (^{235}U) in natural uranium enabled the chain fission reaction to occur over most of the Precambrian period, even in deposits with relatively low uranium concentrations. Natural nuclear reactors may have been a commonplace occurrence on the early Earth.

Many scientists, however, are inclined to believe that the natural nuclear reactors found at Oklo may represent a unique event in the past history of our planet. The most frequent argument is that the accumulation of uranium for reactors was hardly possible. Accumulation is considered to be uniquely the result of oxidation and reduction reactions which required sufficient amounts of organic materials in water and oxygen in the atmosphere. These constituents are considered to have been scarce, if not inexistent, before a significant presence of life.

At present, we have insight into the possible pathways that allowed enough uranium to accumulate and criticality to occur. Moreover, the general characteristics of a typical natural nuclear reactor are reliably established, and the number of reactors that could have been operating on the Precambrian Earth can be estimated.

Uranium Accumulation

Uranium is widely distributed in nature but occurs only at low concentrations. Natural processes for its accumulation are important for the formation of deposits which are sufficiently rich in uranium to make a chain reaction possible. A general belief is that all major deposits of uranium which were formed during the last 2000 million years are the result of a sequence of oxidation and reduction reactions. Initially, tetravalent uranium (U^{4+}) is oxidized to the hexavalent form (U^{6+}), whose compounds are water soluble and can be easily transported to the zones where uranium accumulates. For precipitation to occur, hexavalent uranium must be reduced to the tetravalent state and this can be accomplished in the presence of organic material.

General picture of a Precambian era devoid of organic compounds and oxygen is rapidly and profoundly changing in the light of more recent studies. The latter reveal that organic material of abiotic origin started to accumulate as soon as the Earth's surface became suitable for the formation of organic compounds. These were produced by chemical processes in the atmosphere which were initiated by electric discharges, the action of solar ultraviolet radiation, or the ionizing radiation from radioactive decay. The impact of comets may also have provided larger amounts of organic compounds. Volcanic activity has produced significant quantities of a strongly reducing compound of sulfur, hydrogen sulfide (H_2S), which could have been effective in reducing hexavalent uranium.

Of particular importance are the findings relevant to the appearance of organic material of biotic origin at a time much earlier than has generally been assumed. The isotopic anomaly in heavy carbon is observed in the Isua rocks, and points to the possibility that even 3800 million years ago photosynthetic activity of some kind of living matter may have existed. Furthermore, fossil traces of a complex microbiota have been reliably identified in rocks that are from 3000 to 3500 million years old.

There is also an accumulation of evidence that oxygen was very likely present in small but not insignificant amounts during the span of recorded geological history. A convincing argument for this has been derived from careful analyses of the thorium and uranium contents of well-dated rocks. These two elements appeared very early in the upper mantle and crust, possibly already 100 million years after the formation of the Earth's core. Their concentration ratio should be constant regardless of the rock's age providing their chemical behavior remained unchanged throughout the course of time. However, this constancy does not hold. The thorium to uranium ratio increases from 3 for the oldest known rocks (3800 million years, from Greenland) to 6 for the youngest material analyzed (300 million years, Australia). This trend shows that uranium must have been lost during intensive weathering and sedimentation in the past and that the process already took place at a very early date. The only satisfactory explanation is that oxygen was present. The tetravalent state of thorium does not change and this element is relatively insoluble and remains in the rocks. However, tetravalent uranium is readily oxidized and its water-soluble hexavalent compounds are easily removed by drainage from water. Thus, the thorium to uranium ratio will increase with time.

There exists a process based on intensive chemical and mechanical erosion which requires neither oxygen nor organic matter for the accumulation of uranium. The process may have been particularly effective on the primitive planet because of the presence of acidic gases in the atmosphere and the surface of land devoid of vegetation. An example is furnished by the 3800 million year old Isua rocks, which are chemically precipitated sediments that were deposited by water.

The process in question consists of a cycle of successive steps of weathering-erosion-transportation, and sedimentation. For a nuclear reactor of some 3100 to 4100 million years ago, of which the core was a sphere of 2 meters radius, less than 100 years would have been sufficient to accumulate enough uranium. In such a particular case, a layer of 10 cm thickness had to be eroded from a uranium-deficient deposit (0.4 gram per ton), whose surface area corresponded to 40 square kilometers. The erosion rate is assumed to have been only 1 millimeter per year, which is the present erosion rate of the Alps. Under the intense weathering conditions which prevailed on the primitive planet, the erosion rate may, however, have been substantially higher and, in consequence, the time required for uranium accumulation could have been much shorter than 100 years.

Constitution of a Typical Natural Nuclear Reactor

The principal constituents were several tons of uranium oxide (UO_2) as the fuel, and as much water which served as a neutron moderator and reactor coolant. The total amounts required were dependent on the pertaining age of the Earth, i.e., on the abundance of the fissile ^{235}U. The earliest natural nuclear reactors could have been spheres of only 0.5 meter radius, containing 1.6 tons of uranium oxide and 0.5 cubic meter of water. Considerably larger dimensions and amounts were required when the lower-grade ores were involved or, later, when the abundance of fissile uranium in nature was lower. Then the sphere would have had a radius of up to 3 meters, with 8.5 tons of uranium oxide and 10 cubic meters of water.

The geophysical conditions at the reactor site would have influenced the efficiency of operation and the power produced. In a porous sedimentary deposit, or underneath a shallow body of water, values of temperature and pressure would not have differed greatly from those of the surrounding environment: the water would simmer and at best the power output would be a few watts. At greater depths, equilibrium could have been reached between liquid water and its vapor, thus establishing a supercritical water system resembling that of a contemporary nuclear reactor. These conditions gave rise to higher pressures (hundreds of atmospheres), higher temperatures (hundreds of degrees celsius), and higher power (multikilowatts). It seems reasonable to assume that 1 kilowatt represented an average power output of a typical natural nuclear reactor, which could have produced occasional bursts at a multikilowatt level.

Mixed Radiation and Dosimetry

A nuclear reactor is a source of mixed radiations (particles, electromagnetic rays) of different energies and intensities, the latter depending essentially on the reactor power. A radiation dose is the amount of radiation energy deposited in a given mass of material and is expressed in grays (Chapter 3).

Let us consider the case of a typical natural nuclear reactor with a core radius of 2 meters, which was operating 3200 million years ago at a power of 1 kilowatt. The calculated total dose rate is 47.4 grays per hour. The contributions of the individual constituents of mixed reactor radiation, per hour, are the following: 39.9 grays as fission fragments, 1.2 grays as fission neutrons, 1.5 grays as fission gammas, 3.4 grays due to radiation emitted by the fission products, and 1.4 grays for gamma-rays liberated during the neutron capture reactions.

The dose rate is moderate and similar to that provided by smaller laboratory irradiation facilities. The largest contribution (84 percent) is due to the kinetic energy of fission fragments. Part of this energy is transferred to water and induces the formation of free radicals and subsequent chemical changes related to radiation effects. Most of the energy is localized as heat within the uranium. The reverse is true for penetrating radiation: their contribution to the total dose rate is rather small, but the energy is effectively utilized in the production of chemical changes. Some of the penetrating radiation may escape the reactor core and convey energy to the exterior of the reactor core, for example, to underlying water beds.

When the uranium chain-fission process ends, i.e., during the shutdown period, the reactor core is still an important source of radiation because of the decay of longer-lived radionuclides which accumulated during criticality. In the course of 1 year of operation at a power of 1 kilowatt, as many as 2×10^{21} fission nuclides can be produced. Ten years after the end of criticality, the site is still radioactive, since about 14 percent of the radionuclides have half-lives longer than 1 year. The presence of very long-lived radionuclides will be responsible for some degree of radioactivity even after 1 million years.

The number of fission products that migrate from the reactor core is not high, being about 7 percent in the Oklo deposit, but these are significant as specific, effective, *in situ* irradiation sources. Two of these, ^{90}Sr and ^{137}Cs, circulate in the environment for about 300 years. The total dose released by ^{137}Cs alone may be as high as 4900 million grays; the dose rate will be relatively low because of a long transit time and a large volume of material in which the radionuclides are deposited. Many fissiogenic nuclides have much longer half-lives; one example is ^{129}I, whose radioactivity remains for many millions of years.

How Many Reactors?

So far, only the six fossil reactor cores at Oklo are known. Evidently, much remains to be done before the actual number of natural nuclear reactors can be assessed. In any event, the discovery of a fossil nuclear reactor is not an easy task.

The identification of a fossil reactor site is based on the isotopic anomalies relevant to the depletion of fissile uranium and to the enrichment of certain isotopes which are the end products of the decay of fissiogenic radionuclides. Severe conditions have to be satisfied if the search is to be successful: the geological formation containing the fossil site must have been conserved until the present time; the redistribution of material within the ore must be negligible in order to avoid a dilution effect or a loss of anomalies; the samples for analyses must be taken from the right locations, which requires some luck!

Promising locations are high-grade uranium deposits with at least 20 percent uranium, and which are older than 1000 million years. They should contain 1 cubic meter or more of the uranium-bearing minerals. Finding such places is not a simple matter. It depends on the cooperation of uranium-exploring companies in communicating the confidential characteristics of deposits. Furthermore, as in the Oklo case, reliable identification requires thousands of sophisticated and costly analyses.

Relatively little has been accomplished in the search for other fossil reactors, although there was some activity during the decade following the discovery of the Oklo phenomenon. At that time, dozens of uranium ores were tested for isotopic variations in uranium, neodymium, samarium, and ruthenium. The samples were carefully selected from worldwide locations, although only one sample for each location was analysed (from Canada, South America, Africa, and Australia). The ages varied from 450 million to 1880 million years. No isotopic anomalies of interest were found and it was concluded that criticality had not been reached in the sampling localities. Since the analyzed samples represented only a minor fraction of the potential reactor zones, it was also considered that evidence for remnants of critical operating conditions may still be found, even at these locations.

At the present time, not many uranium deposits older than 2100 million years are known, and the oldest well-dated uranium-bearing sedimentary rocks are between 2500 million and 3000 million years old. The reason for this may lie in their destruction during criticality, i.e., during the operational period of the natural reactors. It was not an atomic-bomb type of explosion that was involved, but rather a burst of energy followed by intense geological activity. These conditions could result in destruction of the reactor core by dispersing several cubic meters of the uranium deposit and "diluting" both the uranium content and the isotopic anomalies.

It is nevertheless possible to make a rough estimate of the number of natural nuclear reactors which existed in the past history of our planet. Reasoning is based on the fact that the ratio of ^{235}U to ^{238}U, which was measured for various deposits on the Earth, is constant to within 0.001 to 0.002. This observation suggests that only 1/1000 to 2/1000 of the total uranium content could have been consumed in natural nuclear reactors. On the basis of this assumption, it is possible to evaluate the total number of natural nuclear reactors in the 1-kilometer-thick layer of the Earth's crust. If only 1/1000 of the total mass of uranium was involved in the chain fission processes, whereby the uranium concentration was 3 grams per ton in the crust and the ratio of fissile to natural uranium was 0.1, it can be calculated that about 100 million reactor sites of the Oklo type could have been active in the past.

Assuming five reactors per site, each operating at an average power of 1 kilowatt for 1 million years, the total energy liberated would be 500,000 terawatt-years. Because of the large time span and the global dispersion, this impressive figure is but a modest contribution to the total energy supplied by other sources during the same period of time (between 4100 and 3100 million years ago).

CHEMICAL AND EARLY BIOCHEMICAL EVOLUTIONS

Scientists have many reasons to believe that living matter originated on our planet under conditions of chemical evolution, a sequence of events that led from simple inorganic molecules to complex organic compounds and eventually to primitive forms of life. Chemical evolution was followed by that of a biochemical type, early biological evolution, which led to the abundance of the early life forms.

A multitude of chemical compounds were available as suitable starting materials for efficient syntheses on our primitive planet. Comparative planetology informs us that the early Earth contained abundant amounts of simple compounds such as water (H_2O), methane (CH_4), oxides of carbon (CO, CO_2), ammonia (NH_3), hydrogen (H_2), hydrogen sulfide (H_2S), and phosphoric acid (H_3PO_4). Studies of phenomena relevant to the early Solar System indicate that collisions with comets could have supplied the young planet with simple organic substances, and not unlikely even some organic molecules which are known to be the building blocks of life, such as amino acids, purines, and pyrimidines (Chapter 6).

Sufficient energy was also available from various sources. Energetic ultraviolet rays induced chemical changes in the atmosphere and in the surface layers of the primitive oceans. Close to the Earth's surface, electric discharges in the atmosphere synthesized a variety of compounds which could have reached the surface waters in unaltered form. The heat of volcanos has been useful both for chemical reactions in the atmosphere and for thermal synthesis on the peripheral lava surfaces.

Of primary importance for efficient chemical syntheses is the specific action of a given type of energy and not its abundance. An example is the Sun, the most powerful energy source, which provides 60 times more energy than all other sources together. However, less than 2 percent of the solar energy (the ultraviolet rays) was directly involved in processes of evolutionary chemical changes. Another example is provided by ionizing radiation. It was a minor partner of the various energy sources which were available on the primitive Earth, but a very efficient one because of its omnipresence and the effectiveness of its free-radical reactions.

The Beginnings

It is not unlikely that life may have existed in Isua time, or at least earlier than 3500 million years ago, but the available data still require futher analysis before a clear-cut answer can be given. Probably the most reliable and definite presently known traces of early life are the stromatolites. These are sedimentary structures formed as a result of metabolic activity of microorganisms, which are recognized from their characteristic wavy layers of carbonate deposits. The oldest stromatolites are found in Australia and Africa in deposits dated at 3500 million years ago. Most likely they are due to blue-green algae (the cyano bacterias), a kind of prokaryotic, bacterium-like microorganism which can release oxygen; numerous strains are also capable of anaerobic (nonoxygen-producing) photosynthesis.

Evidence for such an early appearance of life on Earth is puzzling to some scientists, who consider that several hundred million years are not sufficient for the complex sequence of events called chemical (prebiotic) evolution, and that more time (and chance) was needed before life could begin. There are some suggestions that the earliest forms of life may have reached the Earth from outer space (Chapter 6). Such approaches are perhaps pertinent to the general problem of the origin and existence of life beyond the Earth, but are of limited use in gaining insight into its beginnings on our planet. It is true that a period of 100 million years is so much beyond our grasp that we can hardly appreciate what was probable or improbable within the framework of chemical evolution. However, if we assume that life might have come from somewhere beyond the Earth we are only opening a new file with similar questions (where, when, and how life originated), in addition to other no less delicate problems such as how living species could have survived under adverse conditions of space travel and adapted to the terrestrial environment.

Such approaches divert the search for a direct answer and neglect the impressive achievements of laboratory research. Experiments show clearly that the abundance of energy and chemicals on the primitive Earth must have led efficiently, and on a relatively short time scale (perhaps 1 million years or less), to a wealth of biologically important compounds. A large variety of organic molecules, known as the building blocks of life, have been synthesized by abiotic processes using ultraviolet rays, electric discharges, or heat (to mention only the most common sources of energy used in laboratories). Prebiotic chemistry shows that, starting with simple inorganic compounds as raw materials, the amino acids, peptides, purines and pyrimidines, sugars, and many other biochemical molecules could have been formed abundantly prior to the existence of life on Earth. These compounds could conceivably have participated in chemical processes that produced protocellular structures, the earliest forms of primitive living matter. The progression from random to well-ordered and self-replicating polymers is one of the remaining enigmas.

The sequence(s) of evolutionary events could have occurred in water when our planet became hospitable to organic compounds and chemistry relevant to living matter, i.e., at temperatures between 0 and 100 degrees celsius. These conditions could have prevailed 4000 million years ago, or even earlier. The period of chemical evolution most likely lasted some tens of millions of years. Before early biological evolution prevailed, there must have existed a significant period of time when chemical and biochemical evolutionary processes occurred simultaneously.

Ionizing Radiation as an Energy Source

Estimates of the contribution of ionizing radiation to the energy inventory of the primitive Earth usually take into account only the radioelements dispersed in a 1-kilometer-thick layer of the crust. The contribution of radioactivity to the total energy supply is found to be roughly equivalent to that provided by electric discharges or ultraviolet rays with wavelengths shorter than 150 nanometers. The contribution of natural nuclear reactors to the total energy reaching the primitive Earth's surface was smaller, as can be assessed from the Oklo data multiplied by the estimated 100 million Oklo-type reactor sites.

At present, the oceans cover more than two thirds of the Earth's surface and the area of the sea in the past was even larger when probably only the tops of volcanos were visible in a worldwide ocean. The predominating presence of water, with physico-chemical properties conducive to chemical reactions, certainly conferred a privileged status to the aqueous environment as a setting for chemical evolution. Ionizing radiation was a suitable source of energy for prebiotic chemistry in the primitive aqueous environment.

A nuclear reactor situated beneath the ocean bed would represent a site of abundant radiation chemistry both inside and outside the reactor core. Radiation escaping from the reactor would produce free radicals and new chemical compounds, which could diffuse into the bulk of the ocean water and escape further irradiation and possible degradation. If water (e.g., from underground veins) circulated through the reactor core, irradiation would have taken place inside the core. The fate of synthesized material would depend on the operating conditions of the reactor; normal power variations and the ensuing changes in dose rates and temperature would not be detrimental to the compounds formed. On the other hand, prolonged irradiation together with a rise in temperature during the supercritical stage would render the process more complex because of the destruction of accumulated radiolytic products and the formation of new compounds.

The present chemical composition of the oceans was most likely already established 4100 million years ago. The role of ^{40}K in the chemistry of the primitive oceans is particularly important: since this element was homogeneously dispersed in ocean waters, and delivered about 0.5 megaelectronvolts of energy at each decay, it was a significant source of energy for radiation synthesis. Because of its radioactive decay (half-life, 1.28×10^9 years), its abundance in the Earth's past was about ten times higher than the present 0.01 percent of natural potassium (Table 13). Early ocean water contained 0.38 gram of potassium per liter, of which 0.00038 gram consisted of ^{40}K. From data for the total mass of water (1.7×10^{21} kilograms), its ^{40}K content (6.5×10^{14} kilograms), the rate of radioactive decay, and the energy released, we calculate 2.5×10^{39} electronvolts as the total amount of energy delivered by radioactive decay of ^{40}K over a period of 1000 years. If in this calculation we assume that 0.1 molecule is synthesized per 100 electronvolts of energy absorbed, and that the average molecular weight of radiolytic product is 100 atomic mass units, we find that 1×10^{12} kilograms of material could have been synthesized on the geologically short time scale of only 1000 years.

The chemical action of ionizing radiation may have been important not only for processes in chemical evolution (prebiological), but very likely also for early biological evolution. An analysis of the most significant events in the evolution of the earliest life forms shows that the changes were not primarily of morphological character but, almost exclusively, the result of biochemical changes and metabolic innovations. Such changes could have been effectively initiated by radiation-produced free radicals in the primitive forms of life.

Further Reading

Miller, S.L., and Orgel, L.E., *The Origins of Life on the Earth,* Prentice Hall, Englewood Cliffs, New Jersey, 1974. [A concise, clearly written text that presents in a popular way various aspects of the origin of life on our planet. A book worth reading.]

Rutten, M.G., *The Origins of Life by Natural Causes,* Elesevier, Amsterdam, 1971. [A detailed, scientific presentation offering excellent insight into geological and geochemical aspects. Equally useful for the consideration of problems of radioactivity, chemistry, or biology relating to the study of evolution. Rich in fine illustrations. A book that should be consulted.]

Ozima, M., *The Earth, Its Birth and Growth,* Cambridge University Press, New York, 1981. [A concise, clearly presented role of radioactivity in the past of the Earth.]

Windley, B.F., Ed., *The Early History of Earth,* John Wiley & Sons, New York, 1976. [A collection of papers from a scientific meeting where the Archaean is considered in details. Most of the contributions have the character of review articles and provide detailed information not only on geology and geochemistry of the early planet, but also on the early atmosphere and hydrosphere. Written for scientists, but also of interest to everyone who needs more information on the early stages in the life of our planet.]

Ponnamperuma, C., Ed., *Chemical Evolution on the Early Earth,* Academic Press, New York, 1977. [Scientific papers from a meeting at which the earliest stages in the life of our planet were considered with great courage. They offer detailed information on oxygen and the early atmosphere, as well as on early traces of life. The terminology is primarily intended for specialists, but most of the texts are clearly written and can be used by nonspecialists also.]

Literature on the Oklo phenomenon is scarce. Complete, detailed information appears in two books which were published by the International Atomic Energy Agency of the United Nations. They are collections of the works presented at two international meetings covering all scientific aspects of natural nuclear reactors. The presentation is highly technical, but some of the texts can be read with elementary knowledge in physics, chemistry, and geology:

The Oklo Phenomenon, International Atomic Energy Agency, Vienna, 1975.

The Natural Fission Reactors, International Atomic Energy Agency, Vienna, 1978.

Our planet is permanently bathed in ionizing radiation of both terrestrial and extraterrestrial origins. Radioactivity is omnipresent in nature and almost any sample of rock, water, and air will show its presence if placed in a sufficiently sensitive device for radiation detection. In this respect, the contemporary Earth does not differ substantially from the early planet. Nevertheless, this century is profoundly marked by the discovery of artificial radioactivity and the man-made radioactive substances. Their influence on everyday life is determined primarily by the number of their applications rather than by their abundance.

In this chapter we survey the inventory of radioactive materials and radiation, both natural and man-made, and consider some of their more specific applications; many conventional uses have been indicated elsewhere (Chapters 2 to 4). We examine the radiation exposure of the population, the hazards, and some accidents, and consider radiation protection and techniques employed for safe handling of radioactivity in research and applications.

The storage and disposal of low-level radioactive wastes are also considered; problems related to high-level radioactive wastes and nuclear reactors as well as other aspects of nuclear energy are discussed in Chapter 9.

"Radionuclide" and "radioisotope" are terms routinely used to designate radioactive species; for example, ^{14}C is a radionuclide, an unstable carbon nucleus that undergoes radioactive decay; it is also an isotope (a radioisotope) of the chemical element carbon. There are three categories of radioactive species on the present-day Earth. They are named according to their origins: primordial, cosmogenic, and artificial radionuclides.

We have seen that primordial radioactive nuclides were formed by violent nuclear processes in the stars (Chapter 5), and that, as a result of subsequent circulation of matter, they became incorporated into the primitive Earth about 4500 million years ago. The cosmogenic radionuclides are still being continuously formed by nuclear reactions of cosmic rays with atmospheric and terrestrial matter. Further, artificial radioisotopes are abundantly formed during the operation of nuclear reactors and testing of nuclear weapons, or designed and produced in specially equipped laboratories and industrial plants.

The inventory of ionizing radiation near the Earth's surface consists mainly of radiations emitted during decay of the three categories of radionuclides. The penetrating gamma-radiation dominates and the contributions of alpha and beta-particles remain localized near the site of their origins. The man-made radiation is generated by special machines known as accelerators of charged particles. They constitute a powerful tool in the study of the structure of matter and for some applications of larger doses, but contribute very little to the total amount of ionizing radiation on the Earth. Ionizing radiation from extraterrestrial space, i.e., in the form of cosmic rays, is significant both as a tool in exploring interstellar space and stellar processes, and in its contribution to the total amount of radiation on the Earth's surface.

8

The Contemporary Earth

NATURALLY OCCURRING RADIOACTIVE SUBSTANCES

Primordial Radionuclides

Radioactivity of natural origin arises mainly from primordial radionuclides and their daughter products. These are dispersed in soils and rocks and some of them, after being leached by runoff water, reach surface waters and eventually end up in the oceans. Radioactivity is also introduced in the atmosphere by diffusion of the radioactive gases radon and thoron.

Primordial radionuclides already existed in the matter from which the solar system originated. About 15 primordial radioactive species are presently known in nature and their half-lives vary from 704 million years (^{235}U) to 8×10^{15} years (^{148}Sm).

It is believed that with further improvement of radiation detection methods a few more primordial radionuclides may be discovered among the apparently stable natural isotopes. Their radioactivities are expected to be very low and the intensity of their radiation very weak and difficult to measure with accuracy because of the omnipresent background radiation. This radiation is mainly comprised of cosmic rays, which not only reach every point on the Earth's surface, but also the detectors located in deep caves and mines. Trace amounts of uranium, thorium, and potassium, which are usually present in construction materials, also contribute significantly to the background radiation.

Two chemical elements are of particular importance with respect to radioactivity on the contemporary Earth: uranium and thorium. Their decay creates other radioactive nuclides which, in turn, contribute to what is known as a "radioactive family" or "radioactive series". ^{238}U is the parent of the "uranium family": it has a half-life of 4468 million years and decays with emission of an α-particle. The daughter nuclide is ^{234}Th, which is also radioactive and decays to a radioisotope of the chemical element protactinium, which in turn gives rise to a new radionuclide, and so on until finally a stable nuclide is reached. In this way the ^{238}U series descends through 14 major steps before terminating in a stable isotope of the chemical element lead (Table 14).

The amount of a daughter product in equilibrium with uranium depends on the half-life (Table 15). One ton of uranium contains only milligrams of ^{226}Ra and micrograms of ^{210}Po.

A considerable part of the radioactivity in the Earth's crust is also due to the members of two other radioactive families. One of these is headed by another isotope of uranium, ^{235}U. The parent

Table 14

MAJOR MEMBERS OF THE ^{238}U RADIOACTIVE FAMILY

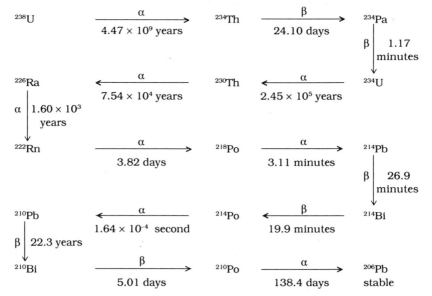

Note: The mode of decay and the half-life of the nuclides are indicated.

Table 15

**AMOUNTS OF RADIOELEMENTS PRESENT AT RADIOACTIVE
EQUILIBRIUM IN ONE TON OF NATURAL URANIUM**

Nuclide	Amount	Nuclide	Amount
Uranium-238	992.9 kg	Actinium-227	0.21 mg
Uranium-235	7.1 kg	Radium-226	340 mg
Uranium-234	54 g	Lead-210	3.77 mg
Protactinium-231	334 mg	Polonium-210	7.4×10^{-6} g
Thorium-230	16 g	Francium-223	4.7×10^{-12} g

nuclide of the third family is an isotope of the element thorium, ^{232}Th, whose half-life is 1.4×10^{10} years.

Uranium

This chemical element is widely distributed in the Earth's crust but rarely occurs localized in significant amounts. In most soils and rocks there are usually only a few grams of uranium per ton.

Because of its use as a fuel in nuclear power plants and in nuclear weapons, uranium has a strategic importance. Its total accessible reserves are very limited; they are estimated at about 5 million tons of uranium in the ores containing 1000 grams or more of uranium per ton. According to more optimistic estimates, the total resources may be three to four times higher than this amount. This is still a modest appraisal in comparison with the extent of uranium accumulated in the oceans, which may be as much as 4000 million tons. However, because of the extreme dilution and costly procedures involved, extraction from this source has so far not proved to be economically feasible. A pilot plant designed for this purpose in Japan was closed in the mid-1980s.

In the human body uranium is concentrated in the brain, heart, and some other organs. The total content found in an average human body is about 100 micrograms but it may be as high as five times this figure depending on diet or, more precisely, on the composition of the soil and fertilizer that are used to produce the basic food. Uranium can be introduced into the soil from fertilizer and is transmitted from thence to food, and subsequently to the human body. Some chemical fertilizers contain significant amounts of uranium because of a relatively high content of this element in phosphates that are used in manufacturing the fertilizer.

Radium and Radon

Radium is the most famous natural radioactive element because of its historical role and its wide use in research and medical therapy between the two World Wars. Some 26 radioisotopes of radium are known, many of which have half-lives so short as a few millionths of a second. The most important isotope is ^{226}Ra, which has a sufficiently long half-life (1600 years) to accumulate significantly in uranium-bearing ores. As seen from Table 14, ^{226}Ra is a member of the uranium family. Before the age of nuclear power, uranium was mined exclusively for the production of radium required to meet the needs of research and medical institutions.

The total quantity of radium that was extracted during the first half of our century amounted to only a few kilograms. It is indeed amazing to realize the enormous impact that such a small amount of one single chemical element has made in numerous fields of human activity, quite apart from the basis thus provided for nuclear physics and chemistry.

Radium is present in all rocks, soils, surface waters, and wells. Being chemically similar to calcium, a chemical element which is fairly abundant in soils, radium is likewise absorbed by plants and is transferred via the food chain to the human organisms, where it becomes concentrated in the bones. The total radioactivity of a human body due to radium and its daughter nuclides is 1 becquerel, i.e., one disintegration per second. It may be as high as tenfold this value in the population of certain areas such as Kerala (India), where the soil is rich in monazite sand and contains significant amounts of uranium, thorium, and their radioactive decay products.

The daughter of ^{226}Ra is a radioactive gas, ^{222}Rn, which has a half-life of 3.8 days. It is a chemically inert substance and readily escapes from the site where it was formed. No place on Earth is free from radon, since the parent nucleus (radium) belongs to a group of chemical elements which are soluble, mobile, and ubiquitously disseminated.

Concentrations of radon may in some cases be very important, such as in poorly ventilated uranium mines. The excavation of uranium ores liberates enormous amounts of radon which would otherwise have remained underground. Observations carried out in the 1940s on workers in Bavarian and Bohemian uranium mines have shown that up to 75 percent of these persons suffered from lung cancer caused by inhalation of radon.

The release of radon increases with surface fissures that appear during geological events such as earthquakes and volcanic eruptions. An activity estimated at 1.1×10^{17} becquerels (3 million curies) of radon was released during the eruption of the Mount Saint Helen Volcano (North America) in 1980. Small amounts of radon are forerunners of larger fissures and are useful in forecasting earthquakes. For this reason radon is permanently monitored in seismic regions.

Radon and radium require that particular precautions be taken in the processing of uranium ores for nuclear energy. These elements are separated from uranium in the early stages of the operation and accumulate in the so called ore-tailings. The amount of radium present at this point may be several orders of magnitude higher than that in ordinary soil, and consequently radon can be profusely released. Fortunately, it is relatively easy, if not cheap, to mitigate the release of radon from tailings by covering the deposits with several meters of soil.

Radon penetrates into the human body by two efficient pathways. Respiration introduces emanation into the lungs and the ingestion of water disseminates the soluble gas throughout the organism. Radon in itself is readily exhaled, being chemically practically inert, but the real hazard comes from the short-lived radionuclides that are produced by its decay. Two of these, radioisotopes of polonium, are particularly dangerous because of the energetic α-particles they emit. Moreover, these radioisotopes are easily fixed by aerosols and dust particles and, once introduced into the body, remain as biologically very harmful and localized sources of radiation.

Thorium and Thoron

Thorium is a chemical element with a geological behavior and distribution similar to that of uranium. It is more abundant, but since its decay rate is lower owing to the much longer half-life, the activities of the two elements in nature are similar.

^{232}Th is the parent nuclide of the thorium family. Its decay products include a radioactive gas, thoron. Like radon, thoron is also a chemically inert substance, but it has a much shorter half-life (only 54 seconds) and, hence, has less time to diffuse from its site of origin.

^{40}K

This radioisotope (half-life 1.28×10^9 years) represents only a minor fraction of natural potassium, 0.0117 percent, but it is a very important primordial radionuclide because of the relatively high abundance of potassium in nature; there are 10 kilograms of potassium in 1 ton of sandstone and 35 kilograms of potassium in 1 ton of granite. It is therefore not surprising that the β-radiation of ^{40}K is the main contributor to heat generated by radioactive substances in the Earth's crust.

One ton of seawater contains 0.38 kilogram of potassium and significant amounts of radiation energy were delivered to the oceans within the geological time scale as a result of the radioactive decay of ^{40}K (Chapter 7).

The total content in the average human is 0.14 kilograms of natural potassium, and ^{40}K is the predominant natural radioactive substance in the body. Its activity is about 3900 becquerels.

Cosmogenic Radionuclides

About 20 radionuclides are permanently produced, mainly by the reactions of secondary cosmic rays with the atomic constituents of the lower atmosphere (nitrogen, oxygen and argon) (Table 16). Their half-lives range from minutes to millions of years, as, for example, for ^{39}Cl (56 minutes) and ^{10}Be (1.6 million years). Radioisotopes of carbon and hydrogen are of particular importance because of the role played by these chemical elements in natural processes relevant to life.

^{14}C is efficiently produced by the reaction of thermal neutrons with ^{14}N and the subsequent release of a proton. It decays with a half-life of 5730 years by emitting a β-radiation (0.156 megaelectronvolts), and is the main contributor to the natural radioactivity of the atmosphere.

Because of its long half-life, radioactive carbon can thoroughly mix with the entire carbon content of the atmosphere, hydro-

Table 16

MAJOR RADIONUCLIDES PRODUCED BY THE REACTION OF COSMIC RAYS WITH ATMOSPHERIC CONSTITUENTS

Nuclide	Half-life years	Mean production rate atoms (per square meter per year)	Total amount on the Earth surface
^{3}H (tritium)	12.33	1.6×10^{11}	7 kg
^{10}Be	1.6×10^6	1.26×10^{10}	430 tons
^{14}C	5730	7×10^{11}	75 tons
^{26}Al	7.2×10^5	4.8×10^7	1.1 ton
^{32}Si	104	5×10^7	2 kg
^{35}S	0.24	4.8×10^8	4.5 g
^{36}Cl	3.01×10^5	5×10^8	15 tons
^{39}Ar	269	4.2×10^{11}	

Note: Data in the two last columns are estimates.

sphere, and biosphere. This is the basis of the so-called "radiocarbon dating", the earliest and still one of the most successful methods for dating objects on the millennia scale (Chapter 2).

The radioactivity of a typical human body due to ^{14}C is about 3700 becquerels, comparable to that of potassium. However, the total energy released annually in the body by ^{14}C is almost ten times lower because of the much lower energy of its β-radiation.

A radioactive isotope of hydrogen, ^{3}H, also called tritium (T), is formed in the upper atmosphere by several nuclear reactions. Its half-life is 12.3 years. The natural production-rate appears rather low, only about 0.5 species per second per square centimeter, but the global inventory is quite impressive: about 3×10^{18} becquerels (80 million curies).

In the atmosphere, tritium is principally bound in water molecules and reaches the surface of the Earth as rain and snow. Tritium is introduced into the body with air, water, and food, but its contribution to the total human dose is low, even compared to carbon, because of the very low energy of its β-radiation (0.019 megaelectronvolt).

RADIATION OF EXTRATERRESTRIAL ORIGIN

From all directions of cosmic space, a steady shower of particles reaches the surface of our planet at speeds which are often close to that of light. These particles comprise primary and secondary cosmic rays and occasionally also solar wind, when the radiation is associated with violent eruptions such as solar flares at the surface of the Sun (Chapter 5).

The dose rate due to the galactic cosmic rays is fairly constant and its value well established. On the other hand, understanding of solar flares is still limited and their appearance cannot be reliably predicted apart from the fact that they tend to appear in 11-year cycles. They erupt suddenly and within a few hours may deliver harmful doses. In some rather rare events, like that of 1972, lethal doses up to 10 sieverts can be attained in 1 day above Earth.

Fortunately, our planet provides natural shielding. The atmospheric blanket ensures efficient protection by slowing down and absorbing many fast particles which arrive from outer space. Moreover, the Earth's magnetic field deflects many of the charged particles, which pass by entirely; others are detracted from the equatorial regions and channeled into scantly populated polar regions.

The terrestrial magnetic field also provides some degree of protection to the crew of spacecrafts orbiting at relatively low altitudes. Soviet cosmonauts who have spent up to a year in space,

with occasional external activities, have showed no visible deteri-
ous effects with respect to health and working capacity.

However, cosmic radiation constitutes a serious risk for crews
of space stations in orbits higher than several hundred kilometers,
during protracted space missions outside the Earth's magnetic
field. It is estimated that the dose received under such conditions
would be 0.5 sievert or more per year due to the galactic cosmic
radiation alone. In the wall-shielded working space of a satellite
orbiting at an altitude of 400 kilometers, the total annual dose of
cosmic radiation would be about 0.32 sievert or more, depending
on the inclination of the vessel's orbit. This is very much higher
than the usual human exposure to cosmic rays, which is only
0.0003 sievert per year at sea level.

A person in a supersonic jet flying at 20 kilometers altitude will
receive 10 to 20 microsieverts (1 to 2 millirems) per hour under
normal circumstances. A giant solar flare (once during the 11-year
cycle of solar activity) could cause an increase of up to 10^4 to 10^5
microsieverts per hour. Flights with conventional aircraft at lower
altitudes lead to 5 microsieverts per hour, but the total dose for a
given distance does not differ greatly since such flights require
longer periods of time. It is noteworthy that the crew of jetliners
making frequent trans- or intercontinental flights may accumulate more than 5 millisieverts per year, i.e., more than the doses
generally permitted for the public at large.

At the surface of our planet, the annual dose equivalent due
to cosmic radiation at sea level is 0.28 millisievert. However, the
effective value depends on the geographical position and increases
with proximity to the poles or with altitude. Figure 27 provides an
overview of the annual dose dependence on altitude.

It is evident that on a vertical line proceeding from sea level, the
increase in annual dose may be considerable. It will be a challenging goal in the planning of future space missions to ensure that
astronauts will not be exposed to disabling or lethal radiation
doses. A shelter protected with at least 9 centimeters of aluminum
(or its equivalent) is recommended for all spacecrafts traveling
outside the magnetosphere. On protracted flights, such as a Mars
mission, all habitable working space should also be shielded with
the equivalent of 7.5 centimeters of aluminum. Further reduction
is not an easy task; an additional 20 centimeters of aluminum
shielding would reduce the dose by less than 20 percent.

MAN-MADE RADIOACTIVE SUBSTANCES

A large variety of radioactive nuclides are produced in minute
amounts in nuclear research laboratories in the form of by-

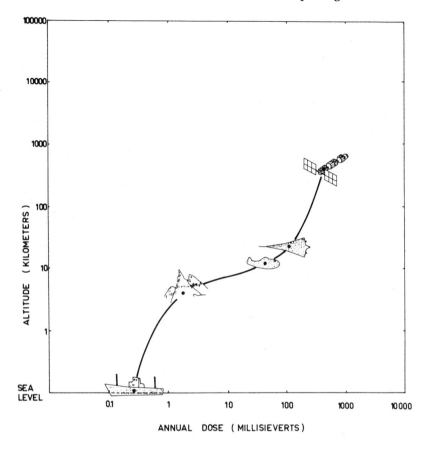

FIGURE 27. Estimate of annual doses due to cosmic radiation. Note the logarithmic scale.

products from the investigation of nuclear processes. Two important categories of artificial radionuclides include those which appear in nuclear power plants from the fission of uranium and other nuclear reactions, and the radionuclides and labeled compounds which are specially produced for applications in various fields of human activity.

Radionuclides in Nuclear Power Plants

Hundreds of various radionuclides are formed during the operation of a nuclear reactor. The fission of uranium alone, which involves splitting of the nucleus in 1 of 30 possible ways, can be a source of about 200 different radionuclides.

The radioactivity which accumulates in a nuclear reactor during operation is enormous. In the core of a reactor with a nominal electric power of 1 gigawatt of electricity (for example, a pressurized water reactor which contains 90 tons of 3 percent enriched uranium), the maximum activity may be as high as 17,250 million curies. The activity of the fission products is preponderant (13,800 million curies), although the contribution of actinides (3450 million curies) and of the radionuclides produced by reactions of neutrons in reactor construction materials (10.6 million curies) is also important.

The presence of short-lived radionuclides will contribute toward a decrease in activity following shutdown of the reactor, but the overall activity still remains very high. In the example given above, there would be a remaining total of 4230 million curies after 1 day, and 963 million curies after 1 month. On a global scale, the total artificial radioactivity on our planet due to the production of electricity by nuclear fission in 1987 was about 4.5×10^{12} curies for a total capacity of 260 gigawatts of electricity.

It is interesting to compare this value with some estimates of global natural radioactivity. For example, 4.5×10^{11} curies represents the total activity of the hydrosphere, since its average activity is 0.01 becquerel per gram and the total mass is 1.7×10^{24} grams. Similarly, the activity of the lithosphere is calculated to be 3.8×10^{14} curies, assuming that the total mass is 2.4×10^{25} g and the specific activity is 0.5 becquerel per gram. These figures show that our planet carries about 4×10^{14} curies in its surface layers down to a depth of 40 kilometers. In contrast, the total contribution of man-made radioactivity corresponded to only 1 percent of this amount in the mid-1980s (Figure 28).

However, if a cosmic event such as collision with a large extraterrestrial body occurred, or if there were a large-scale war (not necessarily nuclear) the role of radioactivity in nuclear power plants could be of primary importance. The reason is that at variance with natural radioactivity, which is dispersed in 10^{19} tons of matter (water, soil and rocks throughout the globe), artificial radioactivity (in 1987) was imbedded only in about 40,000 tons of nuclear fuel, contained in some 400 reactors and located mainly in a few industrialized countries.

Even in the mid-1950s scientists were aware of the problems that could result from the accumulation of radioactive nuclides in nuclear reactors. At that time, the industrial applications of large radiation doses seemed very promising and a number of eminent scientists believed that large amounts of fission products could subsequently be used to make cheap, strong, radiation sources, and thus alleviate at least partly, if not entirely, the production costs of nuclear energy. It happened that the contrary was true. Separation of the bulk of fission products from spent reactor fuel,

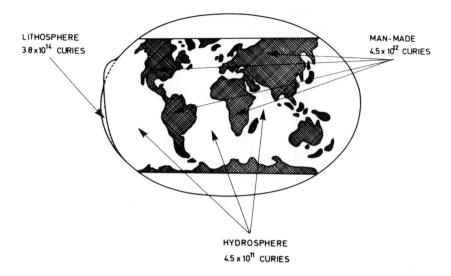

LITHOSPHERE
3.8×10^{14} CURIES

MAN-MADE
4.5×10^{2} CURIES

HYDROSPHERE
4.5×10^{11} CURIES

FIGURE 28. Radioactivity on the contemporary earth. The lithosphere is the outer part of the Earth's surface with a thickness of several kilometers lying upon a mass of hard rock. The hydrosphere is the watery part of the crust including all waters. The artificial radioactivity is that contained in 1987 in nuclear power reactors.

along with the isolation of individual radionuclides and the manufacture of appropriate sources, turned out to be complex and costly.

The Tailor-Made Radionuclides and Labeled Compounds

In 1984, in the United States alone, there were 2.4 million shipments of radioactive materials for practical applications in various fields. The total activity was about 2.5×10^{18} becquerels (68 million curies). In comparison with the extent of radioactivity which was accumulated during the same period in nuclear reactors in the United States, this latter figure was only nominal. With regard to the practical use of radioactivity and radiation, however, it was of paramount importance. For practical purposes a radionuclide must have an appropriate half-life, a suitable form and energy of radiation, and exist in a convenient chemical form. For example, medical diagnosis requires a nuclide with a reasonable short half-life (of the order of minutes). In this case, the utilization of initially high activity is expedient for convenient and accurate measurement, and the danger of subsequent irradiation effects on the organism following examination is limited since the radioactivity decays rapidly. The half-life must be longer (several years) for applications in medical therapy or in technology, where strong external radioactive sources are needed. In tracer work, the use of radionuclides which emit low-energy radiation is preferred because of simpler protection requirements and easier handling of radioactive material.

The chemical form is also important, particularly when radioactive substances are used as tracers of chemical processes of a kind found in agriculture, in an industrial plant, or in numerous human applications. For such purposes, labeled compounds are prepared in which one of the stable atoms in the molecule is substituted by its radioisotope, or by a radionuclide of another chemically similar element.

The "tailoring" of a radioactive substance for specific applications is performed by nuclear- and radiochemists and by scientists trained in the techniques of nuclear reactions and chemical manipulations of radioactive sources. Several hundred radionuclides and thousands of labeled compounds are at present commercially available and routinely used in numerous fields such as medicine, agriculture, geology, or industry.

The commercial production of artificial radioisotopes and labeled compounds is a business which is already 40 years old, and which involves a large variety of products ranging from microgram amounts of transuranium elements (but tons of plutonium!) and millicurie (megabecquerel) quantities of labeled compounds, up to irradiation sources with activities of the order of a million curies. The extensive preparation and use of artificial radionuclides has stimulated industrial-scale production of specific equipment for the safe handling and storage of radioactive materials. It has also promoted the manufacture of robust particle accelerators and special nuclear reactors designed for the production of radionuclides.

SOME UNIQUE THINGS RADIOACTIVE ISOTOPES CAN DO

Nuclear power is still a privilege of economically advanced societies, but radionuclides find wide application both in developed and developing countries because of valuable services which they can provide with reasonably simple and safe techniques. Such nuclides are used in various domains of basic and applied research, as well as in routine activities in medicine, industry, agriculture, or of raw materials exploitation (e.g., mineral extraction and processing).

The applications are based essentially on the consequences of the interaction of radiation with matter: physical (absorption, ionization) or chemical and biological (free radicals, ions, hot atoms). Some of these, already presented in Chapter 3, have been more or less efficient alternatives for nonradioactive techniques in thickness or level measurements, or in the search of defects in welding. Here we consider cases in which ionizing radiation and radionuclides offer a superior if not a unique tool at the present state of knowledge.

Specific Irradiation

About four hundred accelerators of electrons were used in industry in the mid-1980s as sources of radiation for manufacturing goods of improved quality (Chapter 3). Irradiation with low-energy electrons (up to 0.5 megaelectronvolt) is used for the cross-linking of thin plastic films or thin wire insulations. High-power beams (300 to 500 kilowatts) of low-energy electrons enable a high production speed (up to 1000 meters per minute) in the curing of laminating adhesive between wood panels with simultaneous curing of the protective top coat. Electrons in the energy range of 0.5 to 10 megaelectronvolts, with beam powers up to 200 kilowatts, are mainly used for radiation improvement of insulation of wires and cables, and production of heat-shrinkable materials and hot water-resistant polyethylene pipes. Electron accelerators with energies up to 10 megaelectronvolts and a beam power of tens of kilowatts are occasionally also used for sterilization of medical products.

Strong radioactive sources can be employed for intense irradiations without a particular need for an external power supply (Chapter 3). Specially constructed large and powerful irradiation units, which can contain up to several million curies of ^{60}Co, are used for sterilization of medical supplies, drugs or spices. About 145 industrial installations were operating in 40 countries around the world in the mid-1980s, and the total installed capacity was about 80 million curies. At the same time, according to the International Atomic Energy Agency, the annual demand for ^{60}Co was 23 megacuries for medical sterilization and 4 megacuries for sterilization spices and food; the demand in the 1990s is expected to be at least twice as high.

Radiation sterilization appears to be a promising way of combatting detrimental insects and reducing significant losses (15 to 20 percent) of the world agricultural production (both plant and animal). The basic idea is that if a large number of target insect species are bred, sterilized by radiation, and released into the field, the sterile insects will mate with the wild insects. Such mating would not produce offspring, thus leading to a decline in the population which, under suitable conditions, could be reduced to zero. By the end of the 1980s about ten species of pest insects had been successfully handled by radiation sterilization in wide areas (tens of thousands of square kilometers), sometimes in combination with other techniques. One example is the eradication of medfly (Mediterranean fly) in some areas of Japan and Mexico. Radiation sterilization is achieved by exposures of a few minutes in ^{60}Co irradiation units, and the sterilized medflies (tens of millions per load) were delivered by helicopter to the infected areas.

^{60}Co units for medical use require less activity, i.e., several kilocuries. They provide the sufficiently large doses needed in therapy to complete or replace surgical intervention. Short exposures at the position of a tumor at various depths cause only limited damage to skin and tissues which cover the tumor area, and are of great help in cancer therapy. Tens of thousands of ^{60}Co therapeutic units were used as routine equipment in hospitals around the world in the mid-1980s.

Applications in medical therapy also require sources that are very weak and contain only a few millicuries, but which provide radiation energy at the position of a tumor with a minimum degree of irradiation outside the tumor area. Usually, a colloidal suspension of a suitable radionuclide, such as ^{198}Au, is injected directly into cancerous tissues. A colloidal solution differs from a true solution in the larger size of solute particles in the former, thus ensuring that the substance containing the radionuclides remains at the site of injection.

If the radionuclide emits only energetic β-rays, a uniform radiation dose is distributed throughout the injected tissue without significant exposure of other parts of the body. The duration of irradiation and the dose delivered to the tumorous site are regulated by selecting an isotope with a suitable half-life and initial activity. After seven half-lives have elapsed, the radioactivity is reduced to about 1 percent of its initial value as a result of exponential time dependence of decay. This is taken into account in the adjustment of the initial amount of radioactivity, usually at the millicurie level, according to the required radiation dose.

Tracing

Radionuclides used as tracers provide a powerful tool for gaining insight into the course of many processes. The application is based on the fact that all isotopes of a chemical element behave in the same way throughout the course of a physical, chemical, or biological process (with some rare exceptions). Thus, radioactive isotopes which are added to a natural mixture of stable isotopes of a chemical element behave in the same way as other atoms of the same element. However, they have a particular advantage in that the detection of their radiation enables reliable tracing, for example, in a pipeline, in a blood vessel in the human body, or in certain complex steps of metabolism.

Sodium is an important constituent of many chemical processes in the human body. By introducing a minute amount of sodium chloride containing some radioactive sodium into the bloodstream, one can follow directly and externally the course of sodium in the circulatory system and all body tissues. The amount of radioactive sodium used is negligible (10^{-18} gram) and it behaves in the same way as nonradioactive sodium atoms.

In a similar manner, ^{15}O, which emits positrons with a half-life of 2 minutes, can be mixed with air breathed by a patient, thus enabling tests of lung efficiency by radioactive means.

In agriculture, ^{32}P (a β-emitter, half-life 14.3 days) is often used to label a phosphate fertilizer. The optimal conditions for the use of the fertilizer are examined by tracing the β-radiation of phosphorus in plants.

Radioactive tracer techniques are widely applied for the investigation and optimization of mineral processing on a large scale. Such methods are invaluable in hydrology for the determination of the origin and distribution of water resources or the flow of underground waters in large geographical areas.

Radioimmunoassay

Human health depends on a dynamic interplay of thousands of chemical compounds, and for the maintenance of satisfactory standards it may be important to know the concentrations of many of these compounds. This is not always an easy task, since the compounds of interest are generally present in minute concentrations that may be as low as a millionth of a gram. In addition to its high sensitivity, the selected method must also be very specific. It should allow the reliable determination of very small amounts of a compound in the presence of many other similar substances which differ only by the type, or even the position, of a few atoms in the molecule.

Radioimmunoassay (RIA) is such a technique which, besides being sensitive and specific, has the advantage that it can be performed *in vitro* , i.e., outside the patient's body. RIA combines the sensitivity of radiation detection with the specificity of immunology. An example is the determination of hormone in human plasma. Hormones are specific substances which are produced by endocrine glands and regulate various metabolic functions of the organism, for example, in the metabolism of sugar by insulin. They react with antibodies, which are the substances manufactured by the organism to combat foreign substances which enter into the body and interfere with its chemistry; the antibodies react with invaders by combining with them.

An antibody is used as a specific reagent in RIA. The procedure is carried out in the laboratory on a sample of the patient's blood by using a commercially available kit that contains the pure hormone, usually labeled with ^{125}I (half-life 60 days, γ-rays of 0.035 megaelectronvolt), and the corresponding antibody. In the first step, the labeled hormone and the antibody from the kit are mixed to get a complex "hormone-antibody" in which all hormones (labeled with ^{125}I) are bound to the antibodies, and the activity of the complex is determined. In the next step, the blood sample with the unknown content of hormone is added to the "labeled hormone-antibody" complex and the activity measured. Depending on the amount of measured hormone present in the blood (that is not labeled with radioiodine), the activity will be more or less significantly reduced. The decrease is due to the competition of unlabeled and labeled hormone molecules for the antigen: the higher the amount of hormone in the blood the lower the number of radioactive hormone molecules which can bind with the constant amount of antibodies. The amount of hormone present is deduced by comparison of the measured activity in the blood sample with the activity of standards (Figure 29).

Diagnostic Imaging

The use of radionuclides enables more effective imaging of various inaccessible parts of a complex machinery like the human

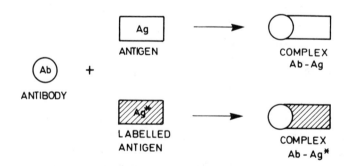

FIGURE 29. Principle of the determination of an antigen by radioimmunoassay. Antigen (Ag) competes with the labeled antigen (Ag*) for a specific antibody (Ab). The amount of unknown Ag is deduced from the activity of the Ag*-Ab complex formed.

body. The principle is based on the fact that a substance which is normally concentrated by a healthy cell of an organ will be depleted in the cells which are damaged by a disease of that organ or replaced by deposits of cancer tissues.

Labeled compounds which are introduced into the body to facilitate imaging must be of pharmaceutic purity. "Radiopharmaceuticals" is the common name for a large variety of substances labeled with radionuclides which are used in the visualization of organs, localization of tumors and the imaging of the dynamics of biochemical processes. The choice of a radiopharmaceutical depends on the organ to be examined and the processes that are investigated; a chemical compound is chosen which is usually concentrated by the organ of interest (Table 17).

Table 17

IMPORTANT RADIONUCLIDES FOR MEDICAL USES

Nuclide	Half-life and decay		Typical applications
^{11}C	20.4 min	β^+	Cerebral blood flow, brain and pancreas imaging
^{13}N	9.97 min	β^+	Lung function test; brain, myocardial and liver imaging
^{15}O	2.03 min	β^+	Lung function test
^{18}F	109.7 min	β^+	Bone scanning, cerebral sugar metabolism, various imagings
^{51}Cr	27.7 d	EC	Measurement of volume of red blood cell
^{67}Ga	78.3 h	EC	Tumor localization
^{99m}Tc	6.0 h	β	Imaging: thyroid, salivary gland, bloodpool, kidney, brain scanning: lung, bone marrow
^{131}I	8.04 d	β	Hyperthyroidism, thyroid cancer therapy
^{133}Xe	5.25 d	β	Ventilation-perfusion measurement

^{99m}Tc is the most frequently used radionuclide for labeling radiopharmaceuticals. It is convenient for human use both because of its radiation and its half-life. It emits γ-rays of 0.14 megaelectronvolt which do not cause excessive irradiation outside the organ where the labeled compound is localized, yet are sufficiently penetrating to be reliably detected. With the half-life of 6 hours enough radioactivity can be concentrated for accurate measurement without long persistence in the body once the imaging is accomplished. ^{99m}Tc is the isomeric nucleus of ^{99}Tc, which does not interfere with the measurements owing to its weak β-radiation and long half-life (210,000 years).

A "technetium generator" permits convenient use of ^{99m}Tc for labeling radiopharmaceuticals despite its fairly short half-life. The device consists of a suitable technical arrangement which provides the short-lived radionuclide in the course of several days. It contains the parent nuclide of technetium, ^{99}Mo, adsorbed on a column of alumina several centimeters in length. Technetium is generated by radioactive decay of ^{99}Mo (half-life 66 hours). It accumulates in the column and can be selectively washed out since it is chemically different from molybdenum. This can be done once or twice a day (within a week) by using a sterile eluent in a simple procedure aptly termed as a "milking" manipulation. The technetium generator consists of a well-shielded and suitable adapted container that enables both safe transportation from the producer and "milking" in the laboratory (Figure 30).

Radiopharmaceuticals are routinely used for the imaging of thyroid, salivary gland, brain, bones, heart, kidneys, liver, spleen, or lungs. Advances in the synthesis of organic compounds and improvements in the methods of radiation measurement steadily increase the number of organs or specific tissues which are accessible to exploration by radiopharmaceuticals. Data can thus be obtained with respect to position, extent of affliction and functional capacity.

Direct, *in vivo* access to the local biochemistry of the human brain is of particular interest in the *in situ* observation of specific changes due to certain diseases. Technically, observations are made by positron emission tomography (PET). This is a procedure which combines the advantages of tomography, a technique that makes possible the imaging of one specific plane of the body, with radiopharmaceuticals labeled with a positron-emitting radionuclide. Following decay, the positrons undergo annihilation with neighboring electrons to produce γ-rays which are subsequently used to generate the image of the investigated site.

An estimated 10,000 "gamma cameras", the imaging instruments used in combination with radioisotopes in medical diagnosis, were installed world-wide in the late 1980s.

Activation Analysis

Minute amounts of chemical elements in a material can be detected by exposing the sample to neutrons. This method is based on the fact that neutrons easily slip into the nuclei of stable atoms and make them radioactive. The newly born radionuclide has a characteristic radiation energy and half-life, which serve as fingerprints for its identification. A milligram or less of the specimen is exposed to neutrons, usually in a specially devised

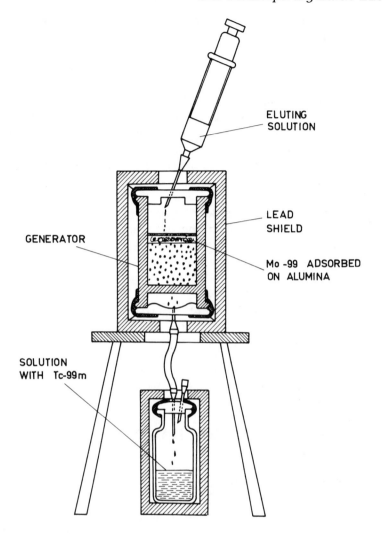

ELUTING
SOLUTION

LEAD
SHIELD

GENERATOR

Mo -99 ADSORBED
ON ALUMINA

SOLUTION
WITH Tc-99m

FIGURE 30. Schematic view of a "milking" device for routine generation of ^{99m}Tc.

channel of a nuclear reactor, and the induced radioactivity is measured after irradiation is carried out. For quantitative determination the activity is compared to that induced in a standard sample containing a known amount of the element. With a neutron flux of 10^{12} per square centimeter per second most elements can be detected in the microgram (10^{-6} gram) to picogram (10^{-12} gram) range.

Activation analysis is routinely employed in the control of ultrapure materials such as those used in nuclear power plants, space technology, or electronics. Since the procedure is nondestructive and efficient for small specimens, it is also widely utilized in the analysis of extraterrestrial materials such as lunar samples, meteorites, micrometeoritic dust excavated from polar ice, and interplanetary dust particles collected by planes flying at high altitudes.

Neutron activation analysis is of great assistance in archaeology; for example, in tracing the circulation of commodities and determining links between different areas of the globe or distant civilizations. The artifacts from various localities of interest in the archaeological study are submitted to neutron activation analysis, radioactivity data compared, and the presence of specific constituents checked.

The technique is occasionally used for chemical analyses in criminology. An example is the detection of the toxic element arsenic in hair, for which the analysis can be made by irradiating as little as one strand of hair. Moreover, since the hair grows at a relatively fixed rate and arsenic enters the hair from the blood into the hair root, the time elapsed between ingestion of arsenic and the effect of poisoning can be inferred from analyses of hair sections.

Where neutrons fail, the procedure can be extended to bombardment with charged particles from an accelerator. A wide choice of particles and energies is available in choosing an appropriate nuclear reaction which leads to a radioactive species that can be reliably measured.

RADIATION EXPOSURE, HAZARDS, AND ACCIDENTS

Omnipresent Radiation

The human species has always been exposed to ionizing radiation from natural sources: cosmic rays, radionuclides incorporated in the body, radioactive substances in the soil or underlying rocks, cosmogenic radionuclides in the atmosphere, and the gases radon and thoron which have already accumulated in caves and to some degree still do in present-day homes. Quite recently the man-made sources of ionizing radiation have been added to the list.

In principle, it is not difficult to evaluate the radiation exposure of a person. It is sufficient to measure the radiation intensities present in the environment and in the body. However, the data will depend on prevailing living conditions such as locality, housing, and diet. They also depend to some extent on the stage of development of the society in which an individual lives and on social norms such as the frequency of special medical controls and treatments. In order to obtain a reliable figure for the average dose exposure it is necessary to conduct a statistical survey of persons working and living under a wide range of conditions.

An Example: Great Britain in the mid-1980s

A survey made by the National Radiological Protection Board of Great Britain shows that in this country a person receives on the average 2.2 millisieverts (equal to 220 millirems) per year, of which 87 percent are due to natural and 13 percent to man-made sources of ionizing radiation. These data will be considered in some detail: they represent average values obtained for a large community for which, besides natural sources, an extended nuclear power program and a significant use of artificial sources of radiation had to be taken into account.

Natural radiation exposures consist mainly of doses received indoors. They arise from inhalation of decay products of radioactive gases which are emanated from materials used in building and construction. The contributions of radon and thoron to the total dose exposure are, respectively, 32 and 5 percent. Depending on the ventilation efficiency of a building, the construction material, its porosity, and the type of wall covering the concentration of these gases can exceed 10 to 100 times the average exposure dose, which is estimated to be 0.8 millisievert per year.

Occasionally the contribution of natural radioactive gases (emanations) may be considerably greater, particularly because of the higher content in some natural waters. For example, the mineral water in Bad Gastein (Austria) is 1 million times more radioactive than water from the public supply. Several decades ago, the visitors of such spas like Bad Gastein were encouraged to drink and bathe in radioactive waters, and even to sit in the "emanatoria" caves and inhale abundant amounts of radon emanating from surrounding rocks. At that time it was believed that the effects were beneficial. From present-day knowledge of the biological action of radiation, such practices would certainly no longer be recommended.

The contribution of terrestrial radioactivity is almost entirely due to penetrating radiation of the three primordial radionuclides ^{40}K, ^{232}Th, and ^{238}U. These provide 19 percent of the total annual exposure dose, 0.4 millisievert. This contribution also depends on the type of soil and rocks, and on the geological history of the region. Some rocks of igneous origin and particularly granites may be up to ten times more radioactive.

Cosmic rays contribute 14 percent to the total dose exposure, with an annual average of 0.30 millisievert. This value varies considerably with latitude and altitude. Near the equator, where the terrestrial magnetic field efficiently reduces the number of cosmic ray particles which reach the Earth's surface, the annual dose is lower (0.2 millisievert). At higher altitudes, such as in some communities in Latin-American countries, the annual doses are much larger, for example, 2 millisieverts at 4000 meters.

About 17 percent of the total dose (0.38 millisievert per year) is due to the radioelements which are incorporated in the human body. Over half of this is due to ^{40}K, which causes nearly uniform irradiation of the body because of the homogeneous distribution of natural potassium. Other contributors include uranium and thorium which accumulate predominantly in the lung and bone cells.

Contribution of artificial sources to the average dose exposure is moderate when compared to natural sources and amounts to only 13 percent of the total with 0.31 millisievert per year. It is essentially due to what is called "medical irradiation".

The term "medical irradiation" should be defined more precisely; it certainly does not represent a general aspect of everyone's environment, even in a society with advanced medical care. The only persons who are irradiated in this way are those who have been submitted to X-ray examination, radiotherapy, or radioactive tests in hospitals. However its consideration as part of the total population exposure is important because of possible genetic and carcinogenic effects. The range of dose exposures in medical applications is large and the doses delivered are not always small. During X-ray examination the body receives 0.05 to 5 millisieverts; in routine diagnosis with radionuclides one is exposed to 0.1 to 5 millisieverts, and in radiotherapy to 3000 to 60,000 millisieverts. These doses are applied locally.

Other irradiations due to artificial sources represent only 1.5 percent of the total dose exposure i.e., 0.03 millisievert per year. They are due to various sources including the global fallout of atmospheric dust particles which incorporate the long-lived radionuclides released mainly during past testing of nuclear weapons in the atmosphere.

Occupational exposure is restricted to a limited part of the population, namely, to persons dealing with radiation in professional activities (industry, medicine, research) or to those working in various fields of nuclear energy.

Figure 31 presents schematically the annual radiation exposure of an individual as derived from data for the population of Great Britain. Catastrophic events can influence the general picture by increasing the contribution from natural sources, e.g., in the case of volcanic eruption, or from artificial sources following inadvertent release of radioactivity from nuclear installations.

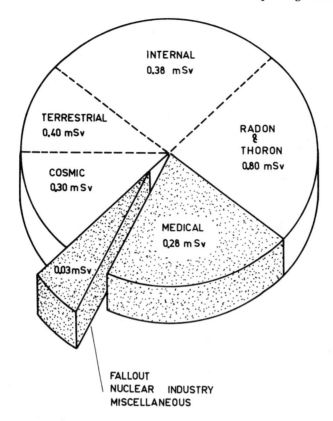

FIGURE 31. Components of the annual dose delivered to a population in Great Britain in 1984. The contribution of natural sources is 87 percent and that of artificial radiation sources is 13 percent. Data from the National Radiological Protection Board. Pie areas are not proportional to dose.

The accident at Chernobyl in 1986 added an extra exposure of 0.03 millisievert during the first 12 months after the accident; the level of radiation then rapidly declined in Great Britain.

The National Radiological Protection Board has made a reassessment of exposure by the end of 1980s. A new estimate of the average exposure within the United Kingdom is 2.5 millisievert per year. The increase is principally due to a re-evaluation of exposure to radon gas and its daughters. Radon indoors accounts for nearly half of the average exposure of the U.K. population to ionizing radiation (1.2 mSv per year).

Individual Dose Exposures

From an understanding of basic aspects of the chemical and

biological action of radiation it is known that harmful effects already exist at the molecular level in the cell (Chapter 3). Even minute amount of absorbed radiation energy induces the formation of reactive free radicals that initiate reactions with biologically important compounds. These reactions result in damage to essential molecules which comprise the building blocks of the cell. Depending on the conditions of irradiation and the individual concerned, the damage can be more or less repaired and the cells can survive and divide. However, there remains a risk of transmission of induced abnormality to the descendant daughter cells.

At higher absorbed doses, such as under accidental conditions or following certain forms of radiotherapy, a large number of cells may be killed before undergoing normal cell division. A lethal dose corresponds to the killing of many cells in the tissues and severe and irreparable damage to various organs. Its level depends not only on the total dose but also on the density of the ionizing radiation (the linear energy transfer, LET), the duration of dose reception (the dose rate), the localization of the radiation effect (some part or the whole body), and, in the case of internal irradiation, also on the rate of uptake and possibility of elimination of the radioactive substances.

For a quarter of a century it was believed that there exists no threshold dose and that deleterious effects of radiation are proportional to the dose absorbed. A linear nonthreshold dose response relationship was accepted. The linear refers to the increments of dose and risk. The acceptable dose exposure is still a matter of conjecture which is formulated as a recommendation and is subjected to revision. Radiation protection guidance is based on the recommendations of the International Commission on Radiological Protection (ICRP). Two sets of guidelines are presently accepted: one for the population at large and another for individuals whose occupation entails exposure to radiation.

In addition to the 2.2 millisieverts, which we have considered as an average for population exposure due to natural and artificial sources, the average radiation dose according to the ICRP (Paris, 1955) should be limited to "a lifetime average annual dose of 1 millisievert"; the overall body irradiation should not exceed 5 millisieverts per year, with the condition that persons exposed to as much as 5 millisieverts in 1 year take care that their accumulated lifetime dose will not exceed the allowed 1 millisievert per year on the average.

Presently an overdose is taken as exposure of the whole body, blood-forming organs, or other critical organs, to 250 mSv or more, and any other external exposure to 750 mSv or more.

For professionally exposed personnel it is recommended that the overall body irradiation not exceed 20 millisieverts per year, while the locally received doses can be larger depending on irradiation conditions and the part of the body concerned.

Exposures to several sieverts (several hundred rems) are lethal. A reassessment of the data on atomic bomb victims in Japan suggest that the lethal dose is likely to be lower than 3 sieverts. Further, the maximum acceptable dose recommended for professionals (20 millisieverts) is not prudent and should be lowered. Some scientists even insist that for professionals the doses should be lowered to levels which correspond to the average annual dose exposure of the population at large.

The exposure to radiation of population must be optimized and kept as low as reasonably achievable, taking into account any relevant economic and social features. In 1988, the European Parliament issued a statement in which limits are set for admissible levels of radioactivity of commercial materials in circulation within the European community. Several examples of these limits are listed in Table 18.

Table 18

**SELECTED REFERENCE LEVELS OF RADIOACTIVITY
(EUROPEAN ECONOMIC COMMUNITY 1987)**

Cesium in milk (level proposed by ecologists)	1 Bq per liter
Natural radioactivity in whisky	50 Bq per liter
Natural radioactivity in milk	80 Bq per liter
Natural radioacticity (^{40}K) in fish	100 Bq per kg
Human ingestion of ^{14}C from food	100 Bq per day
Human ingestion of ^{40}K from food	100 Bq per day
Caesium in milk (proposal from European Parliament)	100 Bq per liter
Radioactivity of potatoes	100—150 Bq per kg
Caesium in milk (proposal by Euratom experts)	4,000 Bq per liter
Caesium in milk (proposal by World Health Organization)	1,800 Bq per liter
Natural radioactivity (^{40}K) in one cubic meter of sea water	12,000 Bq
Radioactivity of ^{137}Cs which corresponds to the tolerated annual dose limit of 5 millisieverts	400,000 Bq

Accidents

The high accidental exposures to ionizing radiation which were recorded in the pioneering days of radioactivity were a consequence of scanty knowledge and experience. At present, such accidents arise more frequently from the negligence of professionals than from lack of knowledge or from technical failures.

It seems that a major risk is presented by portable sources of radiation. Usually, such sources are used in nondestructive examination of metal casting or weldings. They consist of ^{60}Co or

[192]Ir in the form of metal pellets or pencil-like pieces, which are placed inside a lead container. The container provides shielding from penetrating radiation during transportation and storage. It also allows safe manipulation during work when the source is brought into position by remote control. On certain occasions the radioactive material has been left outside the container, or removed deliberately and stored without awareness of danger. These accidents have caused the overexposures and deaths of several persons who have found the source, or of members of their families who were exposed to the unshielded sources in their homes. Documented information about accidents at non-nuclear facilities between 1945 and 1987 reports 53 events in industry, research laboratories and medical centers. They resulted in 156 overexposures and 24 deaths.

Strong radioactive sources used in medical therapy can be very dangerous if the internationally accepted regulations for their use and storage are not strictly followed. Regulations require that the source be delivered only to qualified personnel; moreover, the radiation unit employed must be regularly supervised by experts of a national radiation protection committee. Accidents in Mexico (1983) and Brazil (1987) well illustrate the disastrous consequences which result from failure in observing regulations.

The incident in Mexico began when junkyard workers received as scrap a 3-ton radiotherapy device containing about 400 curies of ^{60}Co. This device, which was never installed and put into operation, was taken from the warehouse of a medical clinic where it had been stored for several years. Two steel foundries in Mexico and one in the United States handled the scrap contaminated with radioactive cobalt. They produced 4000 tons of radioactive steel which was subsequently used to manufacture table legs and reinforcement rods for buildings. The accident was discovered a month later and thousands of inspectors subsequently attempted to trace the contaminated material throughout the United States and Mexico. Opened radioactive cobalt pellets contaminated 20 homes and many persons were exposed to radiation. Most of the doses were low, but over 100 people received significant doses ranging from 10 to 500 millisieverts. Some individuals may have received whole-body irradiation amounting to as much as 15 sieverts, but this occurred over a period of several months.

The accident in Brazil resulted from exposure to ^{137}Cs, also from an abandoned radiation therapy unit that was broken open in a scrap-metal yard. The cesium was in powder form and had an activity of 1400 curies. In all, 244 people were contaminated. More than 20 persons were seriously overexposed to radiation, and four of them had died.

RADIATION PROTECTION

One cannot see, smell, or taste ionizing radiation. The fact that none of the human senses can detect it renders radiation protection difficult, but other factors are also involved. One of these is negligence. A further obstacle lies in the frequently exaggerated fear of radiation that is beginning to prevail in public opinion, mainly because of various controversies on the safety of nuclear installations and radioactive waste disposal. Safety problems often receive large coverage in the mass media, but objectivity may suffer when the hazards are minimized or exaggerated by biased standpoints on nuclear energy.

Principles

Contrary to a general belief, the scientific basis of radiation protection is reasonably well established. It is fair to recall that radiation hazards, although far from being fully understood, are nowadays better assessed than those of most other dangerous agents and pollutants of the environment.

In establishing a reliable radiation protection system it is expedient to call upon several independent, internationally recognized, scientific bodies which follow the achievements in radiation research, especially those relevant to the biosphere and human welfare. These authorities recommend the basic standards for radiation protection which are widely used as the basis for national regulations concerning professional and public sectors. In addition to routine control of national measures for radiological protection, their applications are supervised more and more efficiently by an alarmed public opinion.

It is imperative in radiation protection to reduce as much as possible the dose exposure. The reason is that there is no evidence that the net direct effect of radiation on man is anything but harmful, apart from those medical applications where the benefits are clearly larger than the risks.

Even in medicine the balance between risks and benefits of radiation exposure of patients is not always clearly defined and decisions largely depend on the clinical urgency and the result of diagnosis. The margins are so large from case to case that an appropriate solution often depends on the skill of the clinical staff in avoiding unnecessary doses to body tissue, while still providing a maximum of diagnostic information or therapeutic efficiency.

There are three ways of minimizing the dose from an external source of radiation: reducing the time of exposure, increasing the distance, and providing shielding. A combination of the three possibilities frequently offers the best solution.

When a radioactive material enters the body by inhalation of airborne contamination, ingestion through the mouth, or, in some cases, through the skin or a contaminated wound, protection measures rely mainly on reducing the exposure time, i.e., by applying routine procedures to remove the deleterious substance from the body.

Precautions

There are general rules which must be observed when working with radioactive material. A radiation monitor is used to measure the radiation level which will determine the extent of protection necessary. The instrument also provides a control of radiation during and after the work. Handling of sources of penetrating radiations must be performed behind an adequate form of shielding which usually consists of lead bricks. Many manipulations are not carried in the hand, but with the aid of various types of tongs and specially adapted tools. Even at the lower levels of radioactivity involving kilobecquerels (microcuries) and less, the radioactive material is never handled with bare hands; gloves must be worn in order to avoid contamination of the skin.

Automatic remote-control operations, behind appropriate shielding and with efficient ventilation, are always used in operations involving larger amounts of γ-ray emitters, such as curie amounts of radioactive cobalt or millicurie amounts of α -emitters such as plutonium.

Instrumentation

A large array of both simple and sophisticated instruments is available for the detection and measurement of ionizing radiation. These are used to monitor environmental radioactivity, working areas, and radiation exposures of personnel. In most cases the detectors are based on the well-known properties of radiation to ionize a gas or to fog a photographic film, to mention only two of the most widely used types (Chapter 3).

Ionization chambers are gas-filled containers whose forms depend on a particular radiation monitoring. For example, the pocket ionization chamber is small enough to be worn clipped in the pocket, like a fountain pen. Some even produce audible signals whose frequency is proportional to the dose rate.

Widely used personal monitors are the film badges. A person potentially exposed to ionizing radiation, or handling any appreciable amount of radioactive material, routinely wears a badge-type holder containing film.

Complex monitoring systems in nuclear installations provide continuous information on radiation in the air and in working areas. The monitoring of radioactivity in air, soil, and water in the environment must be carefully performed for efficient radiation protection of workers in nuclear plants and of the population living in their vicinity.

Low-Level Radioactive Wastes

All work with radioactive substances yields radioactive wastes which require proper storage and disposal. The wastes consist of a large range of items and radioactivities. Low-level wastes contain the unused supplies of radioactive isotopes and labeled compounds, or various contaminated items such as protective clothing, surgical dressings, or disposable containers. The high-level radioactive wastes are mainly relevant to nuclear energy programs and are considered in Chapter 9.

A golden rule in dealing with radioactive wastes is that they should not be released into the environment unless dilution or harmless levels can be guaranteed. The concentration of certain elements in biological systems must be avoided, for example, in algae and fish, when surface water is close to the site of disposal.

In most countries the production and application of radioactive substances and sources of ionizing radiation are strictly regulated by laws and authorized only for persons who have qualifications and working permits. Particular attention is paid to collective users such as research laboratories, industrial establishments, and medical institutions, which can be considerable sources of contamination if appropriate measures for safeguarding the environment are not taken.

For all levels of radioactive wastes, radiation protection is achieved by combining the proper conditions of storage and disposal, sometimes after suitable treatment and concentration. The disposal of low-level liquid radioactive wastes is certainly the easiest and cheapest, if it is performed by discharges into large water systems like big rivers or oceans. This aspect is, however, the most controversial and irritating for the public at large as it can also be dangerous if the regulations are not strictly observed.

The transformation of liquid wastes into solids is costly but safer. Such treatment consists of reducing the volume of liquid by evaporation, calcination of dry residue, and its incorporation into concrete, bitumen, or plastics. This procedure minimizes the risk of leakage of radioactive wastes from containers at the site of disposal (ocean beds, abandoned mines, or excavated trenches).

When it is necessary to store solid radioactive materials it is important that their volume be reduced. This is achieved by compressing the material after some previous treatment such as the incineration of organic derivatives or crushing of glass containers.

Further Reading

Friedlander, G., Kennedy, J.W., Macias, E.S., and Miller, J.M., *Nuclear and Radiochemistry*, 3rd ed., John Wiley & Sons, New York, 1981. [Since the appearance of the first edition over 30 years ago, numerous nuclear chemists and radiochemists have grown up with this clearly written, well documented, stimulating book. It is recommended as an introductory text for advanced undergraduate and beginning graduate students who have some background in chemistry but only little in nuclear physics.]

Pochin, E., *Nuclear Radiation; Risks and Benefits*, Clarendon Press, Oxford, 1983. [This book deals with the different forms of ionizing radiation, cosmic rays, X-rays, and the radiation from radionuclides in the environment or in the human body. It presents in popular form an account of the amounts of radiation received from different sources, the kinds and degree of harm that may result, and the available ways of minimizing risks. It is written for readers who are not necessarily familiar with the physical and biological actions of radiation.]

Radiation—Doses, Effects, Risks, United Nations Environment Programme (UNEP), 1983. [A brief survey of facts on natural and artificial sources of radiation, their effects on man, and the subsequent risks. It is for the general public, well written and illustrated.]

Martin, A., and Harbisson, S.A., *An Introduction to Radiation Protection*, 2nd ed., Chapman and Hall, London, 1979. [A comprehensive account of radiation hazards and their control. Assumes no previous knowledge of the subject. Of interest to the general reader, also because of excellent presentation of basic principles of the structure of matter. It includes radioactivity and radiation, as well as the detection and dosimetry of radiation.]

Eisenbud, M., *Environmental Radioactivity from Natural, Industrial, and Military Sources*, 3rd ed., Academic Press, New York, 1987. [Emphasizes the achievements of artificial radioactivity: reactors, fuel reprocessing, weapons, environmental dispersion of radioactive waste products, nuclear waste managements, fallout, nuclear accidents, biological effects of radiation, radiation standards, safety, and risks. Valuable for both the professional and the general reader.]

Hall, E.J., *Radiation and Life*, 2nd ed., Pergamon Press, Elmsford, New York, 1984. [A popular presentation of basic facts relevant to radiation in everyday life.]

We have seen that nuclear energy provides fuel for stars and that nuclear reactors were in operation in uranium deposits within Precambrian Earth. The present chapter deals with man-made devices for the production of nuclear energy, both controlled (reactors) and explosively released (weapons).

In present-day nuclear power plants the most common type of reactor uses uranium as nuclear fuel and water as a neutron moderator and coolant.

Uranium is of primary importance for nuclear reactors. Its processing involves prospection of suitable ores, mining, extraction and concentration enrichment of the fissile isotope (^{235}U), and fabrication of fuel assemblies. Artificial fissile material, in the form of ^{239}Pu, is mainly used for nuclear weapons.

After a certain period of use in the reactor core, uranium fuel elements must be replaced. The spent fuel is highly radioactive and handling during transportation, storage and reprocessing is a delicate, dangerous and costly matter. This is also true for radioactive wastes; some radionuclides have very long half-lives and the disposal has to meet safety standards which apply for thousands of years to come.

A nuclear reactor cannot explode in a way the nuclear bomb does, since its mechanical destruction would precede its nuclear collapse. Nonetheless, accidents may be catastrophic, contaminating the biosphere in large parts of the globe (as Chernobyl, Ukraine), or part of a continent (as in Windscale, Great Britain). Economic disaster may also occur (Three Mile Island, U.S.).

Nuclear weapons are devices containing fissile material (^{235}U or ^{239}Pu) in which, when once triggered, the chain fission process cannot be controlled and the release of energy occurs explosively. When these bombs are supplied with light chemical elements (hydrogen and lithium), the explosive fission of uranium initiates the fusion process of light-atom nuclei, resulting in an additional violent liberation of energy (thermonuclear bomb).

After 4 decades of research and development of nuclear power, an evaluation of the global risks and benefits gives us the following picture: 12 percent of the globally produced electricity, tens of thousands of weapons with nuclear explosives, several hundreds of nuclear submarines and military surface vessels.

9

Nuclear Energy

CONTROLLED RELEASE OF NUCLEAR ENERGY

Chain Fission of ^{235}U

We have already seen how a neutron can penetrate into the nucleus of a ^{235}U atom and lead to its fission into two fragments, thereby releasing more energy (about 200 megaelectronvolts) than in any other nuclear reaction (Chapter 4). Of particular importance is the fact that, on the average, 2.4 neutrons are also emitted and these, in turn, can cause new fissions and liberate further neutrons. This reaction constitutes the uranium chain fission, which is the basis of nuclear energy release for both controlled (nuclear power plants) and explosive purposes (nuclear weapons). A given mass of fissile uranium produces a million times more energy than the same mass of a chemical fuel. For example, 7.5×10^{13} joules is produced per kilogram of ^{235}U as opposed to 4×10^7 joules from 1 kilogram of coal.

If it is assumed that only two neutrons are released in one step, the first nucleus to undergo fission yields about 200 megaelectronvolts. In the second step, an additional 400 megaelectronvolts is released; in the following, 800 megaelectronvolts, and so on. The propagation of the chain process and the accumulation of energy is extraordinarily fast, even though the amount of energy released in each event is relatively small, being only 3.2×10^{-11} watt-second. The reason for this rapid buildup is that the frequency of successive steps is very high and can be 10^{15} events per second or more.

The control of a chain fission process is achieved by governing the generation of neutrons. This is possible because of the emission of "delayed" neutrons. In contrast with the neutrons, which are released within less than 10^{-15} second during the splitting of the uranium nucleus, the delayed neutrons are emitted by various fission products during a period of about 1 second after fission occurs. These products occur only in 1 to 3 percent of the fission events, but they are of primary importance for controlling chain reaction processes in nuclear reactors.

In practice, control is achieved by using a neutron-absorbing material such as boron or cadmium in the form of rods. By varying the position of the control rods in the uranium mass, the buildup of neutrons and the resulting propagation rate of the chain fission can be regulated.

The chain process is accompanied by formation and accumulation of fission products and this leads to a "poisoning" of the fissile material. Some of the nuclides formed such as ^{135}Xe or ^{152}Sm, are efficient neutron absorbers. They act as "neutron poisons" and progressively slow down the chain process rate. In order to maintain and regulate the reaction, the control rods must be gradually removed from the reactor core.

Criticality

The condition for maintenance of uranium chain fission is that, on the average, at least one neutron released in a fission step causes further fission. This will largely depend on the extent of loss of neutrons by leakage through the surface of a uranium reactor assembly. The smaller the amount of uranium, the larger the surface-to-volume ratio and the probability of neutron escape. A minimum amount of fissile material is therefore needed for maintaining fission. This is the so-called critical mass.

The total amount of uranium required for criticality depends on the degree of enrichment in fissile isotope. It may be as small as a few kilograms if pure ^{235}U is available, but up to several tons in the case of only slightly enriched uranium (with 2 to 3 percent of fissile material). Criticality can also be achieved with natural uranium with only 0.7 percent of fissile isotope, but this requires both a large amount of uranium and heavy water as a neutron moderator.

The fission neutrons are fast neutrons since they are rich in energy (up to 10 megaelectronvolts) and travel at high speeds. A highly enriched fissile material is required to achieve criticality for a fast neutron fission process. The slow or thermal neutrons penetrate more easily than fast neutrons into fissile nuclei and the fission probability of ^{235} U is about 400 times higher than in the case of fast neutrons (Table 11, Chapter 4).

The slowing down of the fast neutrons released in fission is accomplished by maintaining the fissile material in a moderator. The velocity is decreased by the neutron striking the nuclei of the moderator atoms. It is important in these collisions that loss of neutrons be kept as low as possible, i.e., the moderator must have a low absorption cross section for neutrons. The moderator consists of light atoms such as hydrogen, deuterium or carbon and in the first two instances ordinary or heavy water may be used. The thermalization of a fission neutron requires about 20 collisions with hydrogen in water at room temperature, and about 120 with carbon atoms in graphite. A fission chain reaction in uranium is schematically represented in Figure 32.

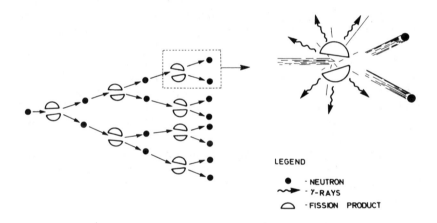

LEGEND

● - NEUTRON
〰 - γ-RAYS
△ - FISSION PRODUCT

FIGURE 32. The fission chain reaction.

Nuclear Power Plants

In a conventional power station, heat is produced by the combustion of a fossil fuel, i.e., by a chemical reaction. The heat is removed by a coolant that directly or indirectly generates steam to drive turbines, which in turn produce electricity. In a nuclear power plant the operations are similar, but in this case heat is produced by the nuclear reaction.

Thus, there are certain similarities between the two ways of generating electricity. In the nuclear reactor neutrons are responsible for the propagation of processes which are generating heat, whereas in conventional plants oxygen molecules participate in chemical combustion. In both cases fluxes have to be controlled and losses minimized to ensure an economic use of fuel.

Moreover, the generation of electricity produces wastes. Those from a nuclear power plant are radioactive and consist of fission products and actinides in the spent fuel, together with various solid, liquid, and gaseous materials which are formed during operation or reprocessing of the reactor fuel elements. The operational wastes from a fossil fuel plant are ashes, combustion gases, fly ash, and radioactive effluents as well. It is not generally realized that coal contains primordial radionuclides in significant amounts. When coal is burnt, the radioactive gases radon and thoron are released, and nonvolatile radioelements such as uranium and thorium, together with the products of their radioactive decays, become concentrated in the ashes.

Heat produced in the fission process is mainly in the form of kinetic energy of the fission fragments, but there is also some contribution from radioactive decays and the prompt neutrons. In order to avoid melting of the uranium fuel assembly an efficient means of heat removal must be provided. The coolant must have suitable heat conducting properties, display adequate resistance to radiation, and be a weak neutron absorber. Coolants which are presently in routine use include ordinary and heavy water and the gases helium and carbon dioxide. Sodium, a soft and silvery white metal which melts at 98 degrees celsius, is used as a coolant in fast neutron reactors in which large amounts of heat are released in a small volume.

At present, nuclear energy production is based mainly on the thermal neutron fission process. The contribution of power reactors using fast neutrons was less than 1 percent at the end of the 1980s.

COMMON TYPES OF NUCLEAR POWER REACTORS

At the beginning of the 1990s the most common reactor type in nuclear power production is the "pressurized water reactor" (61 percent) and the "boiling water reactor" (23 percent). Most of the other types were run only to meet specific needs or in an attempt to introduce them to the market.

Pressurized Water Reactors (PWR)

We have already stated that water can be used as a neutron moderator and coolant. In the PWR it circulates through the reactor core under a pressure of about 150 atmospheres, hence the name applied. The overpressure prevents water from boiling at 320 degrees celsius, which is the temperature inside the reactor core. This superheated water circulates in a closed system, called the "primary" circuit, from the reactor core to a steam generator and from there back to the core. Heat transferred to this generator is used for production of steam in a "secondary" circuit, which drives the turbine producing electricity (Figure 33).

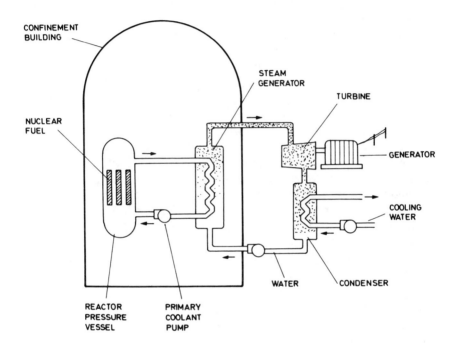

FIGURE 33. Diagram of a pressurized water reactor (PWR).

Water in the primary circuit contains radionuclides formed by neutron capture in impurities, or released from fuel elements in the event of failure. In principle, there should be no possibility of radioactive contamination of the water used to drive the turbines because there is no direct contact between the water in the two circuits. In practice, however, contamination can (and sometimes indeed does) occur because of leaks in the conduits. Therefore, the entire steam supply system (reactor and steam generator) is enclosed in a confinement building. The primary function of this is, of course, to retain any radioelement that may be released from the reactor or from the primary water circuit. The building is constructed with concrete walls and a steel liner. These are sufficiently thick to shield the operating personnel from radiation, and strong enough to withstand the rupture of the pressure vessel, but walls of this type are inadequate in the event of a chemical explosion such as that which occurred in the Chernobyl accident. The pressure within the confinement building is kept at a level lower than that of the outside air. In this way, any inadvertant leakage of radioelements toward the exterior is prevented. The air inside the building is pumped out through a filtering system and is eventually released to the atmosphere through a tall stack.

When ordinary water is used as moderator and coolant, as is generally the case, the reactor fuel consists of uranium oxide with 2.2 to 3.5 percent of ^{235}U. The uranium load is 80 to 100 tons for a nuclear power plant operating at 1 gigawatt. The fuel is sealed in a zirconium alloy, and the other structural materials of the fuel assemblies and core are made of stainless steel.

In a limited number of installations, heavy water, D_2O, is used in a PWR; this material has the advantage of low neutron absorption and permits the use of nonenriched uranium as reactor fuel. This makes the fabrication and reprocessing of nuclear fuel considerably cheaper. On the other hand, the production of heavy water is an energy-consuming and costly procedure. Furthermore, with natural uranium the power density is low and the core size must be large in order to provide an adequate output. This, in turn, requires greater amounts of heavy water and a corresponding increase in the investment costs.

Boiling Water Reactors (BWR)

In this type of reactor, ordinary water is used for neutron moderation and cooling in the same way as in the PWR. The main difference is the suppression of the intermediate heat exchanger and of the secondary circuit.

The water used for cooling the reactor circulates through the core and is converted into steam within the reactor vessel. This steam drives a turbine which generates electricity; the condensed water is then recycled back to the reactor core. The fuel load is similar to that of a PWR.

It is evident that the BWR offers a higher degree of efficiency in power production. However, the steam which runs the turbines is contaminated by radioelements which are formed and accumulated during the repeated passages of the water through the reactor core. For this reason, the turbine must be sealed in a leakproof casing, which also collects the steam losses and circulates them back to the reactor. The turbine hall is a radiation-controlled area and thus all technical maintenance requires special precautions.

Graphite as Moderator

The graphite reactor was popular in the early days of nuclear energy because of relative ease in manufacturing its components. There was no need for uranium enrichment or for heavy water. In the first generation of nuclear reactors, the fuel consisted of metallic uranium canned in a magnesium alloy. The coolant was carbon dioxide gas at a temperature of 400 degrees celsius at the outlet.

This setup also had its disadvantages. Natural uranium provides a relatively low neutron flux and the assembly must be large in order to minimize the escape of neutrons. Moreover, a considerable amount of energy is consumed in circulating the cooling gas through the reactor core. Improvements were made by using graphite-moderated, water-cooled nuclear reactors which were fueled with slightly enriched uranium.

In principle the nuclear power reactors in Chernobyl, as in many Soviet stations, belong to this category. Uranium enriched with 2 to 3 percent of ^{235}U is used, and the cooling water flows under pressure through separate conduits which surround the fuel. The water removes heat from fuel elements, boils, and generates steam which, before being conducted to the turbine, must be collected from thousands of pressure pipes. Although the water effectively takes up heat from the fuel elements, the graphite moderator also collects heat and has to be cooled separately. In order to prevent overheating of the graphite and the risk of fire, a separate cooling system circulates a gaseous mixture of helium and nitrogen.

Reactor for Plutonium Production

Plutonium requirements for military purposes was the reason for the operation of a significant part of the world nuclear reactors within the last decade.

Weapons-grade plutonium must contain less than 7 percent of the most common impurity, plutonium-240. Its amount depends on how long the plutonium was "burnt" in the reactor and can be up to 19 percent. This is why military reactors operate at exactly the right rate to produce the isotopic mixture of plutonium that is needed for weapons. The most common fuel in these reactors is uranium of natural isotopic composition, which provides both the fissile isotope ^{235}U and the breeding material ^{238}U. The moderator consists of graphite or heavy water, and the coolant is ordinary water or a gaseous material.

Training and Research Reactors

For training and research low power units are sufficient. When designed for use at universities, they are of simple conception and are relatively inexpensive. Enriched uranium is contained in aluminum cladding and ordinary water serves as moderator and coolant.

Experimental reactors are designed primarily for research, and frequently play the role of a pilot plant for larger scale projects.

According to the International Atomic Energy Agency, by the end of the 1980s in 55 countries 326 nuclear reactors were used as a training tool, to support analytical studies, and to produce radioisotopes used in medicine, industry, and agriculture.

Fast Breeders

If fast neutrons are used to build up a self-sustaining chain reaction, more plutonium may be produced than is required to enrich the fuel in the core. The moderator is absent and more neutrons are available than in the fission caused by thermal neutrons. By surrounding the core with a blanket of natural or depleted uranium, the excess neutrons are absorbed by ^{238}U to form plutonium. The uranium is chemically processed to recover the plutonium which is subsequently used to manufacture new reactor fuel. This is the concept of "breeding" the "fertile" ^{238}U to make fissile plutonium.

However, the probability of neutron interaction with a heavy nucleus decreases significantly when the neutron energy increases and the uranium fuel employed must be highly enriched with ^{239}Pu (or ^{235}U, which is more expensive). High concentrations of fissile material cause a large number of fission events in a small volume, and the problem of heat removal becomes delicate. A technical complication lies in the fact that core dimensions in fast nuclear reactors are much smaller than those in corresponding thermal neutron units with the same power rating. The coolant must therefore have a very high heat capacity and thermal conductivity, and these properties can only be provided by liquid metals.

Sodium has excellent thermodynamic properties and appears to be the best choice as a coolant despite serious drawbacks. It is highly inflammable, and must be kept strictly isolated from water and oxygen. Moreover, sodium becomes highly radioactive during its circulation through the reactor core. A secondary circuit, also containing liquid sodium, is used to transfer the heat from the primary radioactive sodium circuit to the steam generator and circuit driving the turbine. Figure 34 is a simplified representation of a fast breeder reactor (FBR) for power production and the breeding of plutonium.

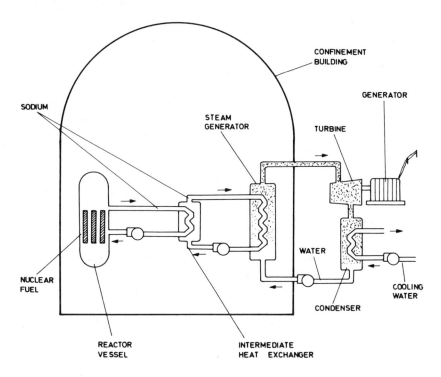

FIGURE 34. Diagram of a fast-breeder reactor (FBR).

It is worth mentioning that only a few fast breeders are in operation at the present time. What makes fast breeders scarce when they seem so promising? One reason is that current needs in electricity can still be satisfied with fossil fuel such as coal, or with uranium extracted at reasonable cost from ores. Other reasons are more technological such as efficient cooling of small volumes at unusually high heat densities, the necessity of using the dangerous material sodium as coolant, and the hazards associated with the extremely toxic plutonium as a fuel.

The breeding of plutonium is in itself not the most important argument in favor of fast breeders. Other types of reactors may be even more efficient in this respect: a PWR provides twice as much, and a graphite-moderated, gas-cooled reactor four times as much plutonium. The principal advantage lies in the conversion of depleted uranium into plutonium. Large amounts of uranium depleted in ^{235}U appear in the tailings of reactor fuel reprocessing, or following isotopic enrichment, and are at present considered as a waste material. For example, about 20,000 tons of depleted uranium was stored near a small enrichment plant in Great Britain in the course of recent years. This material could be used for breeding enough plutonium to produce an energy equivalent to 50,000 million tons of coal.

URANIUM FUEL CYCLE

In terms of the total amounts of material handled and the requirements for manpower, land surface, and resources, uranium fuel production is a small industry compared to that of the fossil fuels. The annual production at the beginning of the last decade was estimated at 40,000 tons; by the end of the century it may reach 100,000 tons of uranium oxide if the expected global increase in nuclear power production takes place.

Occurrence

Uranium is rather widely distributed in nature but represents on the average only between 2 and 4 grams per ton of the Earth's crust. Most of the known uranium resources of the world are found in a few geologically restricted areas, of which the most important are in Africa and North America. The currently mined ores contain from 0.35 to 30 kilograms of uranium per ton. The element is also recovered as a by-product of gold and copper mining, and from the phosphate and phosphoric acid industries.

Extraction

After the ore is mined, a simple chemical procedure is used for the concentration of uranium. A yellow material called "yellow cake" is obtained which contains at least 65 percent uranium.

Tailings from the extraction have practically the same volume as the feed supply of the ore because of the small uranium content. At present, there is a store of 120 million tons which is located mainly in North America. If the current trends continue, it is expected that by the end of this century there will be at least 500 million tons of accumulated tailing material. Precautions must be taken in the storage of these residues because they contain nearly all of the naturally occurring radioactive daughters from the decay of uranium, especially ^{230}Th with a half-life of about 80,000 years and ^{226}Ra with a half-life of 1600 years. In addition, large amounts of radon can escape if the material is not properly confined (Chapter 8).

Various processes are used to produce high-purity uranium from the yellow cake, either in preparing it for direct manufacture of nuclear fuel elements based on natural uranium or in producing fuel charges with enriched uranium.

Enrichment

In natural uranium the fissile isotope content is only 0.72 percent, and for most nuclear energy purposes it has to be enriched. Two procedures are currently in use and these are based on the slight difference in the atomic mass of the two isotopes.

One method involves gaseous diffusion, which was developed and used during the Second World War. It is based on preferential molecular flow of the lighter isotope, ^{235}U, through the micropores of a membrane. Since the difference in the flow rates is small and the separation yield of each step is very low, a large number of separation units must be arranged in a "cascade". In a typical low enrichment plant there are 1000 to 1500 stages which provide uranium enriched to a few percent; in this form, it can be used in reactors moderated by ordinary water. The energy required in this process is high and amounts to 10 percent of that which the enriched uranium will provide as a fuel. In the procedure a highly poisonous gaseous compound of uranium, namely, the hexafluoride, is used. This substance is also chemically very reactive and its handling requires special precautions. The expenditures connected with the operation of a gaseous diffusion process are high and, on the global scale, only a few industrial plants are now in operation.

Another procedure, based on centrifugation, is of secondary importance for the present production of enriched uranium. The separation of the uranium isotopes, also in the form of their hexafluoride derivatives, is carried out in a strong centrifugal field inside a rotating cylinder. The single step enrichment is much greater than with gaseous diffusion, but is still insufficient, and cascades are also required in this case.

Of more recent date is a photochemical procedure which uses a laser beam for dissociation of the gaseous uranium compound, and also relies on the isotopic effect. It provides a fairly high degree of separation in a single step. The realization of an industrial plant is beset with numerous technical difficulties, but several pilot plants were under development in the U.S., Japan, and Europe.

Fuel

Uranium dioxide, manufactured as ceramic pellets, is used as fuel in most nuclear power reactors. The advantages of ceramics over metal are that they undergo fewer dimensional changes at high temperatures; further, they are less easily damaged by radiation, are chemically more inert, and are effective in retaining the gaseous fission products.

The pellets of ceramic uranium oxide are loaded into cans of zirconium alloy or stainless steel to give fuel pins, or elements. The requirements of the canning material are very stringent: it must be chemically inert towards the coolant, should have high resistance to the action of radiation, and provide optimal thermal contact.

The first step involves the fabrication of uranium oxide pellets. The pellets are produced by sintering, a procedure in which the material is compressed under heat at a temperature below the melting point. This method yields a coherent mass of the required density. Subsequently, each pellet is precision-ground to a close dimensional tolerance. The whole process is carefully controlled, since the microscopic structure of uranium oxide pellets is of primary importance in determining the life and performance of the fuel in the reactor.

The number of pellets required depends on the type and power of the reactor. The fully loaded core of a PWR designed to operate at 1 gigawatt of electric power contains about 12 million pellets, which make up a total of about 100 tons of uranium. Usually 200 pellets are placed in one fuel pin, and 300 fuel pins make one fuel assembly. There are 180 fuel assemblies in the reactor core. The spent fuel assemblies are replaced with fresh units, usually one third at a time, at intervals depending on the power and operating condition of the reactor. In normal operation this occurs about once a year.

Several important stages in the processing of uranium from ores to nuclear fuel are represented schematically in Figure 35.

ARTIFICIAL FISSILE NUCLIDES

^{239}Pu

When a ^{238}U nucleus absorbs a neutron, ^{239}U is formed which decays by two successive beta (β)-emissions to the long-lived

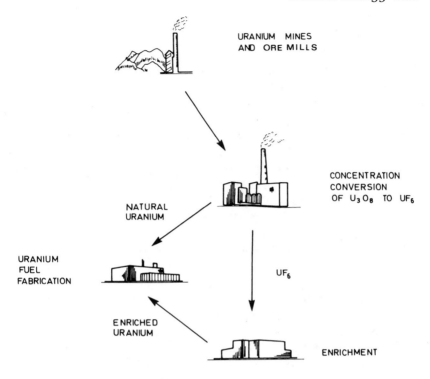

FIGURE 35. The uranium fuel cycle.

nuclide ^{239}Pu, as shown below:

$$n + {}^{238}U \xrightarrow{\gamma} {}^{239}U \xrightarrow[23.5\ \text{min}]{\beta^-} {}^{239}Np \xrightarrow[2.35\ \text{days}]{\beta^-} {}^{239}Pu \xrightarrow[24,100\ \text{years}]{\alpha}$$

During the World War II nuclear reactors were developed and constructed for the production of plutonium for military purposes. Traces of plutonium are also found in nature, in uranium-bearing strata. They are formed by the same reaction with cosmic ray neutrons, and with neutrons emitted in the spontaneous fission of ^{238}U.

Plutonium is technically and economically more readily accessible than ^{235}U. Its present use is mainly for nuclear weapons and it may have a future as a fuel in fast breeder reactors. It has also been considered as a fuel for light water reactors, as a blend of plutonium and uranium oxides. In that case, plutonium oxide will become a commercial item on the market. By the year 2000, France, Great Britain, Germany, and Japan are planning to produce up to 25 tons annually from reprocessed nuclear reactor fuel. Depending on the extent of global nuclear disarmament, it is also expected that plutonium from military stocks will appear on the market.

The present world production of plutonium is not precisely known because of secrecy in its military use and in the operation of reactors for plutonium production. About 1.5 tons of plutonium is assumed to be the annual production of military reactors at Savannah River (U.S.). The oldest of four reactors supplying plutonium for weapons in the United States was shut down after the Chernobyl accident. This reactor has a graphite-moderated and water-cooled core and no concrete containment dome; it has been producing about 0.6 tons of plutonium annually for many years.

The reprocessing of spent reactor fuel is not a common practice, and only a minor part of the plutonium formed in nuclear electric power plants or in various experimental reactors is actually being extracted. However, rough estimates of plutonium accumulation can be reliably made. To begin with, about 0.4 gram of plutonium is formed daily in PWR operating at 1 megawatt(thermal). In the mid-1980s, about 220,000 megawatts (electrical) was available in nuclear power stations, which corresponds to about 740,000 megawatts(thermal). If a reactor operates on an average of 250 days per year, the overall energy produced amounts to 1.85×10^8 megawatt(th)-days, and subsequently 74 tons of plutonium is formed globally per year. The total amount of plutonium which has been accumulated in recent years from the generation of nuclear electricity throughout the world is very likely 6 to 7 times higher.

The entire amount of this element present on Earth, whether in stocks of military deposits or in spent nuclear fuel, is uncertain. According to some authorities the stock of plutonium in the United States at the end of 1984 was between 86 and 100 tons. Judging from reports on the number of Soviet nuclear warheads, reserves in that direction may be comparable.

Because of the very long half-life and extraordinarily high radiotoxicity of plutonium, which is retained in lungs, liver, and bones, these stocks are potentially a great hazard for man and the environment. The handling and storage of large amounts of this element are also dangerous because of the risk of criticality and the unleashing of an uncontrolled chain fission process.

The manufacture of plutonium fuel requires very stringent precautions throughout all operations. The production of plutonium oxide powder, pellets, and fuel assemblies, must be carried out by remote control. Dust levels must be strictly controlled and maintained at low limits. Glove boxes are used for various manipulations to avoid direct contact between personnel and the radioactive material. As in the production area itself, these boxes are maintained at subatmospheric pressure. In the event of a leakage the outside air will flow inward and prevent escape of plutonium from the working area.

^{233}U

A third practicable fissile nuclide is ^{233}U. It is bred from thorium exposed to neutrons in a nuclear reactor according to the following sequence:

$$n + {}^{232}\text{Th} \xrightarrow{\gamma} {}^{233}\text{Th} \xrightarrow[\text{22.2 min}]{\beta^-} {}^{233}\text{Pa} \xrightarrow[\text{27.0 days}]{\beta^-} {}^{233}\text{U} \xrightarrow[\text{160,000 years}]{\alpha}$$

This reaction presents two attractive practical aspects: the element thorium is five times more abundant in the Earth's crust than uranium, and ^{232}Th is its natural form.

^{233}U may be a promising reactor fuel in the future. At present, however, the technology of thorium is in an initial stage and extensive research and development are still required.

SPENT NUCLEAR FUEL

The useful life of nuclear fuel in a reactor is limited for several reasons: depletion of fissile nuclides, accumulation of fission products that act as neutron poisons, mechanical deterioration, and radiation damage of the cladding and the fuel itself. The latter has to be replaced each year, usually to the extent of one third of its amount, despite the fact that the total amount of consumed uranium fuel is scarcely several percent.

The fuel load varies with the type and size of the reactor. It is of the order of 100 tons for a pressurized water reactor for a power output of 1000 megawatts of electricity, and consists of 96.7 tons of ^{238}U and 3.3 tons of ^{235}U. After 3 years of operation at full power, about 5 tons of reactor fuel will be consumed, of which 2.4 tons is ^{238}U and 2.5 tons ^{235}U. The burning of such an amount of nuclear fuel produces enough electricity to meet the demands of an industrial city with a population of 1 million inhabitants for 3 years.

Storage

The handling and storage of spent fuel require special techniques because of the high radioactivity and intense generation of radiogenic heat.

After removal from the reactor and 10 days of "cooling" to allow the decay of short-lived radionuclides, 100 tons of spent fuel still generate heat at a power of 60,000 kilowatts, which will decrease to 1000 kilowatts after 1 year. In the same time span, the amount of activity passes from about 10,000 million curies or more to about 100 million curies. Thereafter, the decrease in heat and radioactivity becomes slow on the scale of a human lifetime. It takes a century before the generation of heat diminishes to about 40 kilowatts and the activity to about 5 million curies. Reduction of these values to a few percent of the initial level would require a period of several thousand years (Figure 36).

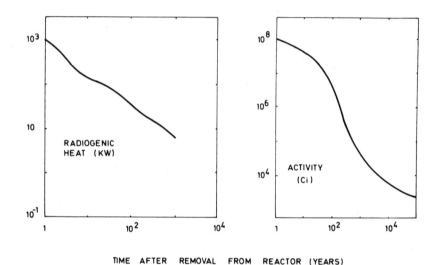

TIME   AFTER   REMOVAL   FROM   REACTOR   (YEARS)

FIGURE 36. Decay of radiogenic heat and radioactivity in 100 tons of spent fuel of a 1-gigawatt PWR power plant.

The storage of spent nuclear fuel is costly because of the need for permanent cooling and efficient shielding against radiation. Some producers of nuclear power prefer to transfer the spent fuel and the rights pertaining there hereto to another country for reprocessing or storage.

Spent reactor fuel is usually stored in water pools at the reactor site or in a repository some distance away. It appears that water provides the best conditions for both shielding and cooling, and there are no problems with respect to further corrosion of the fuel cladding. In order to increase the storage capacity and minimize the risks of renewed criticality, fuel assemblies are stored close together in compact racks and protected by alternate layers of a neutron absorbing material. In this way, a typical storage capacity for a light water reactor such as a PWR can contain the spent fuel from 5 to 10 years of reactor operation.

Various other concepts are considered and are occasionally used on a pilot-plant scale; these include the use of a dry air-cooled caisson, concrete casks, or geological (hard-rock) sites.

Reprocessing

Reprocessing is an important aspect of nuclear energy production because of technical, political and military implications. The aim of treating spent fuel is to separate the radioactive wastes in order to facilitate their treatment for storage, as well as to recover the components which can be used again in reactors or for other purposes. At present, such processing is economically unfeasible because of the relatively low cost of uranium.

Only about 10 percent of the spent reactor fuel from nuclear electric power stations has been reprocessed in recent years. In the course of several decades before 1980, the reprocessing plant in Selafield, Great Britain, treated a total of only 25,000 tons of spent fuel. In general, it seems that for most nuclear plants the solution of interim storage has been adopted pending decisions for reprocessing or disposal in repositories.

Reprocessing begins in the pond, where the fuel is mechanically chopped into small pieces prior to dissolution in nitric acid. Three solvent cycles are used to extract uranium and plutonium. In the first cycle, highly radioactive fission products are removed from the feed stream containing the two elements. The latter are subsequently separated into individual streams and purified and concentrated in the second and third extraction cycles.

Effective reprocessing imposes several general requirements. The recuperation of fissile material must be maintained at a very high level of efficiency for economical reasons. Because of the intense radiations, particularly in the early stages, the process is carried out entirely by remote control, within cells which are heavily shielded and completely enclosed. Any accidental release of radioactive material must be avoided and stringent monitoring of the working space and environment is necessary.

An essential aspect of the process is that inadvertant attainment of a critical mass of fissile material be obviated at all times and stages of the operation. This is particularly important in the reprocessing of spent fuel from fast breeders, where high concentrations of plutonium are inevitably present. Reprocessing plants of this type are still in the course of development and they will have to be more elaborate than plants which treat the spent uranium fuel from thermal neutron reactors. In addition to avoiding criticality, they must handle much larger amounts of radioactive fission products, more intense radiation, and much greater amounts of heat.

At each phase of reprocessing, waste products are generated. Air used for ventilation during the initial stages involving chopping and dissolving contains two gaseous β-emitters with long half-lives: ^{129}I (16 million years) and ^{85}Kr (10.7 years). Before being released, the air is treated so that at least the iodine remains absorbed. Recent regulations require that the krypton be trapped also. Occasional alerts with respect to public safety, particularly in areas where reprocessing plants operate, are mainly due to accidental release of these two radioelements into the atmosphere. The discharge of liquid wastes from reprocessing plants, particularly at Windscale, Great Britain, between 1975 and 1979, was severely criticized. It occurred because of defective technology and laxity in regulations.

At the present time, technical and economic considerations indicate that a commercial reprocessing plant should have an optimum capacity of about 1500 tons of spent fuel per year. The fact that the existing commercial plants in various countries generally do not meet this production level appears to have political rather than economic grounds. The total capacity of reprocessing plants operating throughout the world is not precisely known, since the main suppliers are still military reactors for plutonium production.

For 100 tons of reactor fuel, the spent material contains 3.5 tons of various fission products. The latter are the main source of radiation and heat, and their treatment is a necessary step toward the safe and definite disposal of wastes over a geological time scale. The spent fuel also contains precious nuclear materials such as the residual, unburnt, fissile ^{235}U (0.8 ton), the fissile ^{239}Pu (0.8 ton), and fertile uranium that can be introduced into the manufacture of reactor fuel (94.3 tons). In addition, there remains about half a ton of alpha (α)-emitting radioelements that do not exist in nature and which are of considerable scientific and technical interest. Thus, the initial 100 tons of fuel leads to 4 kilograms ^{244}Cm (half-life 18.1 years), 12 kilograms of ^{243}Am (7370 years), 50 kilograms ^{237}Np (2.1 million years), and up to 460 kilograms of ^{236}U (23.7 million years).

RADIOACTIVE WASTES AND NUCLEAR ENERGY

At the present time there appears to be no urgency in the reprocessing of all spent reactor fuel. The responsible authorities of nuclear power plants have no problems with the supply of uranium and are not particularly interested in plutonium. Moreover, the total amount of spent fuel corresponds to less than 10,000 tons per year on a global scale, and this amount can be accomodated under conditions of controlled storage. Even in countries in which reprocessing plants are operating, only a small fraction of spent fuel is actually being recovered, and the high-level wastes are stored mainly in liquid form in stainless steel tanks.

Because of the harmful effects of radiation on the biosphere and the longevity of certain radionuclides, the problems of nuclear wastes are of interest to both laymen and professionals who are concerned with environmental protection. There is an increasing tendency to oppose the continuous development of nuclear power, not only in the public at large, but also in a number of technical and scientific communities. This is apparent even in countries like France or Sweden, where nuclear electricity contributes efficaciously to the energy supply. Following a national referendum in Sweden, no new nuclear power plants will be built and the existing ones will be completely shut down by the year 2010 at the latest.

The public still seems to be concerned mainly with problems relevant to spent nuclear fuel and to the wastes from reprocessing plants, in particular with respect to fission products. It is, however, not generally realized that radioactive effluents and wastes are generated at all stages of nuclear energy production. These wastes appear during the mining and milling of uranium ores, in the manufacturing of uranium fuel, and during the operation, maintenance, and decommissioning of nuclear power plants.

Depending on their origin, the wastes differ with regard to the nature and amount of radionuclides, physical state, and volume. Like many other pollutants of the present-day environment, they are dangerous to varying degrees and require different kinds of treatment. The management of nuclear wastes is now an important field of research and development in nuclear technology. The hazards must at least be reduced, even if they cannot always be eliminated.

Sources and Amounts

Considerable volumes of wastes are produced during uranium mining and milling, and appear in the mill "tailings". For example, in order to obtain sufficient uranium for 1 year of operation in a 1000-megawatt plant, 120,000 tons of a typical ore containing 0.2 percent uranium must be mined and milled. The volume of the tailing is 58,000 cubic meters , which can be confined within a surface of 1.5 hectares. The total activity is about 700 curies, of which 10 percent comprises the long-lived ^{230}Th with 80,000 years half-life and the 1600 years of ^{226}Ra. Radon will be continuously released for several millennia, and the deposit may have a potential long-term radiological impact on the environment. Precautions are required to prevent contact with the public and to reduce the dispersion of radionuclides, in particular the escape of radon into the atmosphere and the leaching of radium into ground and surface waters.

The wastes which accumulate during the isotopic enrichment of uranium and the fabrication of fuel contain relatively small amounts of natural radioactive elements and do not present particular problems for the environment.

A significant amount of waste is generated during the maintenance and operation of a nuclear plant. Residues from maintenance, repairs, and replacements of reactor components are bulky, but not highly radioactive. In principle, fission products are retained within the fuel, but in practice, even during normal operation, part of the fuel cladding may become defective and allow seepage into the cooling system. The coolant may also contain fission products from extraneous uranium on the surface of the cladding, together with various radionuclides produced by neutron activation of structural materials and corrosion products.

The contaminated water from a nuclear reactor may spread to various parts of the power plant by leakage, and the radioactive substances can appear in liquid waste streams. For removal of these contaminants from the main bulk of the water, there exist well established techniques such as ion exchange, precipitation, filtering, or a combination of these. Of course, these procedures generate their own wastes (radioactive ion-exchange resins or precipitates) which may require further treatment before disposal.

In addition to iodine and krypton, xenon and tritium may also appear as gaseous radioactive wastes. Their amounts are insignificant in the normal operation of a PWR because the primary coolant system is independent of the stream which drives the turbines.

The storage of spent reactor fuel also produces radioactive wastes which arise from contaminated surfaces or as a result of deficiencies in fuel cladding. Such contaminants may thus appear in the water or in the air-cooling systems.

A breakdown of the equipment in reprocessing plants is a particularly serious problem because contamination will be high and voluminous.

High-level liquid wastes are stored in stainless steel tanks 50 to 5000 cubic meters in capacity, enclosed in concrete cells. Radiogenic heat buildup in tanks may be of the order of dozens of kilowatts per cubic meter, and water cooling must be provided with the aid of coils installed in the storage tank. Control of eventual leakage is afforded by a monitoring system at the storage site.

Solid reprocessing wastes consist of cladding hulls, filters and resin beds, noncombustible rubbish, or incinerator ash. The combustible wastes include ion exchange resins or filters. After calcination they are converted into a form suitable for transport and disposal, for example by incorporation of the ashes into a glass melt which becomes a vitrified material on cooling.

Wastes which contain significant amounts of radionuclides that emit α-rays, or high-energy β- and γ-rays, and have high radiotoxiciy and heat output are called the "high-level" wastes. As an example, those from the first stage of the fuel reprocessing operation contain the nongaseous fission products, traces of unseparated plutonium, and other transuranium elements. Most of the radioactivity of the fuel is contained in this fraction.

Storage and Disposal

There is an essential difference between storage and disposal. Radioactive waste left in "storage" is accessible for inspection, recovery or repacking at some appropriate time. In case of disposal, there is no intention of retrieval. The disposal site is not necessarily abandoned and may be subject to surveillance according to international regulations.

The problems of storage and disposal of nuclear fuel wastes concern other users of nuclear energy as well as the services which process spent reactor fuel. The country that provides reprocessing facilities may return the radioactive wastes to the supplier, who will be responsible for their storage on his own territory.

Despite criticism and protest, the disposal of low- and intermediate-level wastes in the sea is still a current practice. Since 1975 it is regulated by an international convention on marine pollution, which requires that national authorities control the burial procedures in the same manner as for disposal on land, and that limitations with respect to the type of nuclides and the activity level be respected. According to this convention, the coastal zones are to be used only for disposal of low-level wastes, while the site for other types must be located at depths greater than 4000 meters.

In principle, the disposal of high-level nuclear wastes must ensure that within 100,000 to 1 million years no significant amount of radioactive material will reach the biosphere. Such a task is evidently very demanding, both from the point of view of technology and ethics. It is the responsibility of the present technological generation to take adequate measures to protect not only itself, but also future generations.

National nuclear energy authorities around the world are searching for appropriate solutions to waste disposal. These activities are at present limited to research in laboratories and pilot plant installations. According to the United Nations Scientific Committee on the Effects of Atomic Radiation (UNSCEAR), there is at the present time no established method of disposal for high-level radioactive wastes on an industrial scale. In France, annual amounts of about 800 tons of such wastes have been vitrified and stored in air-cooled vaults in the course of recent years.

The Search for Long-Term Solutions

Various procedures have been devised to incorporate low- and intermediate-level radioactive wastes into matrices of concrete, bitumen, or plastic, since these materials are resistant to leaching. Disposal in this form can be assigned to shallow ground repositories, abandoned mines, or excavated rock cavities. Complex geological, hydrological, and climatological studies must be carried out prior to selection of a burial site.

The site for long-term disposal of high-level wastes must be at an acceptable distance from areas of probable tectonic activity such as earthquakes and volcanism. Possible contact between the waste and circulating underground water must be avoided. The host rock formation, together with the surrounding geological and hydrogeological environment, should possess natural barriers which serve to prevent or retard a migration of radionuclides from the repository to the biosphere. Computer models for risk assessment also take account of climatic changes.

A better solution for providing long-term safety is a deep geological repository, such as in a mine at 500 to 1000 meters below the surface (Figure 37). The packaged wastes are transported to tunnel rooms and subsequently placed in predrilled holes. Holes filled with waste are then sealed. After all holes in one room are occupied, the room itself is filled with drums of conditioned wastes and then isolated from the remaining depositories. Finally, the entire space is sealed. Similar deep disposal options are: a depository under the sea bed with access from land, or from an off-shore structure.

The realization of this kind of mine repository appears to be feasible in the light of existing experience. There remains however, the problem of adequate packaging materials, which must withstand corrosion and leaching for a duration of 100,000 years. A further problem is how to label the site in a readily understandable manner for distant future generations.

Various other concepts for the disposal of high level wastes have been considered. One of these, which has met with some

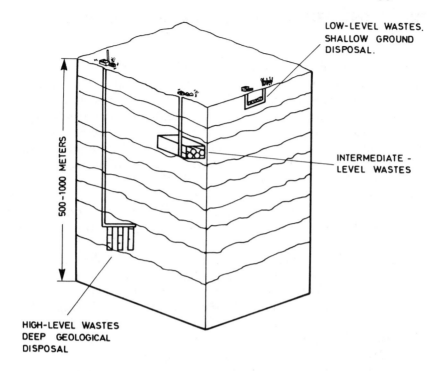

FIGURE 37. A geological repository for nuclear wastes.

degree of public approval, consists in sending wastes into space where they would be located in a distant solar orbit or even outside the solar system. This is hardly feasible at the present state of our technology, and would involve undue risks and costs. Another proposal involves pumping liquid wastes to a depth of 500 to 5000 meters, into porous fractured strata that are known to be suitably isolated.

There exists a fundamental approach to the problem that may provide a technical solution. At present, only three components are separated from the spent fuel: uranium, plutonium and a remaining fraction which includes all fission products, transuranium elements and radionuclides from cladding and reactor material. The storage of high-level wastes could be greatly facilitated if efficient separation of the transuranium elements from the bulk waste is achieved. Neptunium, americium, curium, and plutonium are all long-lived α-emitters which constitute a major potential hazard on a time scale greater than 1000 years. Once separated, the transuranium elements could be consumed in nuclear reactors and, at the same time, contribute to energy production as part of the fissile material.

DECOMMISSIONING OF A NUCLEAR POWER PLANT

Like any other industrial plant, nuclear power plants become obsolete or simply wear out and have to be put out of service permanently. This operation is called "decommissioning" and consists in closing down a nuclear facility in a manner that provides adequate protection to the workers, the public, and the environment. By the end of the present century, decommissioning will affect at least 100 nuclear power plants which are now in operation. About 30 years is considered to be a reasonable lifetime for a power reactor, of which the first were put into service in the 1960s. In other instances, maintenance costs have become exorbitant or, following an accident, the expenses for repairs and renewed operation are too high.

The ultimate goal of decommissioning is a complete dismantling of the plant and a cleanup of the site so that it can be eventually released for unrestricted use. Optimistic estimates indicate that the costs of decommissioning for a 1000-megawatt-of-electricity plant are about 10 percent of the value of a new plant. For a plant damaged by accident, the costs may be several times higher, provided that dismantling is indeed possible.

Apart from funds, decommissioning requires adequate equipment, well-trained personnel, and an appropriate area of land for waste storage and disposal. Decommissioning is therefore a complex task, particularly in developing countries. A less drastic alternative might be to seal off the most radioactive components (e.g., the reactor core) and decontaminate and remove the less active parts. On a short-term basis, it would be possible to remove only the easily accessible radioactive parts, leaving the bulky machinery and inaccessible radioactive components intact.

Experience gained so far in this respect is not extensive, and apart from accident cases such as at Windscale, Three Mile Island, and Chernobyl, has been obtained mainly through the dismantling of low-power reactors and small-scale reprocessing plants. Decommissioning has not yet reached the status of routine industrial activity.

The matter of dismantling a nuclear plant is much more complex than merely putting an industrial establishment out of service, although in both cases much of the work is done by a giant wrecking ball and crane. Operations involving radiation risks are admittedly limited to the reactor core and its components, the water used as reactor core coolant, and the heat exchangers and their piping. However, this aspect dominates the whole operation. It requires a carefully planned approach, thorough radiological protection, and, as much as possible, the use of robots. It is also the most expensive part of the operation, mainly owing to large amounts of bulky, radioactive materials.

The reactor vessel contains most of the radioactivity; once the fuel assembly and the water coolant are removed, radiation hazards are no longer a major obstacle. Most of the remaining radioactive material (about 95 percent), is located inside or immediately around the reactor vessel. This activity is due essentially to surface contamination by radioactive dust, and this can be removed by brushing and with high-pressure water jets. The resulting washings are radioactive and must be treated in the same way as liquid wastes. Concrete and other construction materials which were exposed to neutrons are radioactive and chemical cleaning and mechanical scraping are necessary. Special precautions are required in these operations to prevent the spread of dust and fumes which contain radioactive materials, and all operations involving brushing or cutting are performed on components that are submerged in water.

These and similar operations will benefit from experience acquired in the use of robots and manipulators in the nuclear field. Robots can be programmed, trained, and retrained for each type of new activity. Manipulators have many features in common with those of a robot, but can be operated directly, usually manually and frequently by remote control. New solutions in robotics and automatics appear promising for efficient and cost-saving operations in the decommissioning of nuclear installations.

SAFETY AND ACCIDENTS

The core of a nuclear reactor is a hot place both because of energy released during the chain fission process and radiogenic heat. Radionuclides accumulate during reactor operation, and in a 1000-megawatt-of-electricity PWR 10,000 million curies are already produced after a short period of criticality; heat is generated to the extent of about 300 kilowatts per liter.

Nuclear reactors are designed and constructed to ensure efficient heat removal and minimize the release of radioactivity. They are fully adapted to normal conditions of operation and have a safety margin for certain emergencies which may arise.

Possible Failures

The overheating of fuel elements may occur under various accidental conditions. Transient events increase the reactor power beyond the capacity of the cooling system. A decrease in the coolant flow or, worse, a rupture in the cooling system can cause a sudden and drastic reduction in heat-removal efficiency. In such cases an automatic shutdown of the reactor reduces the generation of heat, while a rapid injection of water from storage tanks assures emergency cooling without the intervention of operators.

Heat generation in the core continues even after the shutdown of the reactor. This is due to energy liberated by the decay of radionuclides, that is, radiogenic heat. Immediately after shutdown, this energy is equivalent to about one tenth of the full reactor power, or about 30 kilowatts per liter in the example given above. Production of heat falls rapidly with time because of decays of short-lived radionuclides. However, cooling must be maintained for the long-lived radionuclides, which continue to generate large amounts of heat.

Reactor design must also take into account the release of radioactivity which will occur in the event of damage to the fuel cladding. Of major concern is a potential escape of larger amounts of radioelements if the fuel elements melt from overheating. Several barriers are provided to prevent such escape. The first is the hermetic cladding of the uranium fuel. A second barrier is the cooling system, which is designed to retain the radioactivity in case of a leak in the cladding. Should further leaks occur, a confinement is foreseen, or some special encasing that will prevent radioactivity from being dispersed into the environment. This container consists of a leak-tight steel shell that may be either independent or in the form of a lining within the concrete building. Originally, it was not designed to withstand pressure created by chemical explosions, but after the Chernobyl accident this aspect is also being considered in the design of new reactor buildings.

There is a popular misconception that a nuclear reactor can explode like an atomic bomb. This fear is supported by reports concerning the amount of fissile material in the reactor core, which is as large as that contained in several hundred nuclear weapons. Further, it is also known that the handling of fissile material in reprocessing plants or in storage areas must be performed with care in order to avoid an accidental assembly of critical mass, which could cause a surge of heat and radiation and a release of fission products into the environment. Such accidents have indeed occurred. In nuclear reactors a state of uncontrolled criticality leads faster to the melting and mechanical destruction of the reactor core than to an explosive liberation of nuclear energy. The reason for this is that the neutron multiplication in the reactor core is too slow to produce an incident of the latter kind. Before the critical mass of uranium in the core can blow apart, about 50 grams of fissile material would have to be fissioned within 1 microsecond; such conditions would then produce an explosion equivalent to 1 kiloton of trinitrotoluene (TNT). This cannot be achieved except in special setups such as those employed in nuclear weapons.

Major Accidents

Documented information about accidents at nuclear facilities with significant overexposure to radiation reports 27 events between 1945 and 1987. They resulted in 272 overexposures and 35 deaths.

Quite a few accidents in the past, however, have remained under cover of military or commercial secrecy, owing to competition between manufacturers, or because of prestige and national pride. It required 26 years to publish the full story of the large accident in 1957 at Windscale (now Sellafield) in Great Britain. The situation has improved under the pressure of public opinion, and there was a relatively rapid and detailed release of information on the more recent and major accidents, at Three Mile Island (U.S., 1979) and Chernobyl (U.S.S.R., 1986).

The Windscale accident occurred in a nuclear reactor that was designed and used solely for plutonium production. It used natural uranium as fuel, graphite as a moderator, and was cooled by air circulation. The origin of the accident lay in a peculiar type of damage induced by neutrons in graphite. Under the effect of irradiation the graphite swells, its thermal conductivity decreases, and the moderator accumulates thermal energy. In order to reestablish the original properties of the graphite, the stored energy must be released, and this is usually accomplished by gently heating the graphite moderator after the reactor has been shut down. The manipulation is routinely performed by restarting the reactor at a low output level and carefully controlling the rise in power and the degree of cooling. The increase in heat permits the release of energy stored in the graphite. It seems that in some parts of the Windscale reactor the cooling was not adequate, with the result that the graphite became hot and the uranium fuel elements began to burn. The fire was subsequently transferred to the nearby graphite. Within 2 days, a large amount of volatile fission products was released into the atmosphere and deposited over wide areas of England and in some parts of Europe. The fire was extinguished on the fifth day by flooding the reactor with water, and the whole reactor system was then sealed off.

It is estimated that during the accident large amounts of radioactive noble gases and about 20,000 curies of ^{131}I, were released into the atmosphere from the 135-meter-high stack. Surveys of radioactivity in the countryside indicated that the highest level of γ-radiation was 40 microsieverts, i.e., 4 millirems, per hour. The contamination of the soil was very extensive within an area of 500 square kilometers; the activity of ^{131}I was between 3700 and 37,000 becquerels or 0.1 to 1 microcurie per square meter. The use of milk from the restricted area was prohibited for more than a month.

A report by the Medical Research Council Committee of Great Britain was published soon after the accident, and concluded that there were no deleterious effects to the health of workers in the Windscale plant or the general public. However, a document published 26 years later, in 1983, by the Radiological Protection Board of Great Britain clearly showed that serious consequences resulted from the accident. Analytical data were provided for the radioactivity released, the meterological conditions, and the contamination of soil and the population. One of the conclusions indicated that only ^{131}I was responsible for 260 cases of thyroid cancer, of which 13 are expected to be fatal. The contribution of other fission products to the total harmful effects was estimated to be four times smaller. There is still controversy about other harmful consequences of the Windscale accident, particularly with respect to leukemia and other forms of cancer among the population living in the vicinity.

The accident at Three Mile Island occurred in one of the two reactors which generate electricity. The reactor core, of PWR type, started to heat up like the heating element of an electric kettle that had boiled dry. A simplified picture is that the accident was initiated by a mechanical failure and that the operators did the rest: in the confusion that followed the automatic shutdown of the reactor, and which lasted several hours, they did wrong things at the wrong time.

In this accident, it is true that the 3.7 meter high reactor core was at no time really uncovered in the same way as a dry kettle. But the water level had fallen to about 1 meter for probably 2 hours. Several more hours were needed to reestablish cooling by natural circulation after the core had been refilled with water. While the cooling was interrupted, radiogenic heat, mainly due to decay of fission products, accumulated and caused overheating of the fuel elements. At some points the temperature in the reactor core was close to 3000 degrees celsius and some fuel elements melted; the melting point of uranium oxide is 2815 degrees celsius and that of its zirconium cladding is 1982 degrees celsius. The core's container did not melt despite its much lower melting point, indicating that most of the intense heat was confined to the central region of the reaction core.

During the accident, a valve was stuck open and the multistory building was doused with water and steam carrying fission products. The steam pervaded the walls, floors, and equipment. About 4000 cubic meters of radioactive water filled the confinement and an auxiliary building, while 2500 cubic meters filled the basement of the reactor building. The water contained a significant fraction of radionuclides which had accumulated in the reactor and included noble gases (50 percent), iodine (40 percent), cesium (20 to 40 percent), and barium and strontium (0.5 to 1 percent). However, since no explosion occurred and the confinement building was not damaged, the radioactive substances did not, on the whole, escape into the environment; there was, however, some release of noble gases (5 percent) and traces of iodine.

Water from the building was pumped out and its radioactivity content removed by passing it through ion exchangers. The building remained highly radioactive and was accessible only to remote television cameras and robots. Together with the robots that had been designed for the nuclear weapons industry, personnel equipped with special protective clothing did the work of cleaning up the building.

Up to 900 workers were involved in different stages of this task, a greater number than that needed to run and maintain the reactor. And all this for an installation that will never run again. On various occasions the question was raised: why not seal up the building and walk away? However, this apparently simple possibility would have turned the reactor into a high-level radioactive waste repository. Neither the site nor the local population was prepared for an entombment which would last for millennia, burying a content of 1000 million curies undergoing gradual decay (Figure 36).

The decommissioning process of the burnt-out reactor was still in progress in 1987, and the final assessment of the accident was not available until 8 years after it happened. However, it was clear that in addition to the expensive and time-consuming cleaning operations, the accident also had important economical and psychological consequences.

This accident is an example of how nuclear energy production can have disastrous economic consequences even without directly endangering human life and the environment. In addition to dismantling costs, that are comparable with those of constructing a new reactor, there are investment losses for the power station as a whole; the second nuclear reactor of the station was not damaged during the accident but it was also closed, thereby incurring further financial loss.

It is generally believed that the greatest damage was suffered not by the business that owns the power station, but by the nuclear energy industry in the United States and elsewhere. The accident showed that despite all precautions and safety measures in design and construction, costly failures can occur even in nuclear reactors of reputedly efficient and safe operation.

The impact on the nuclear electricity industry was disastrous. In the years following the accident not a single order for a new nuclear plant was registered in the United States. Moreover, many orders were cancelled, and numerous reactors that were under construction or ones that had already been built were not put into operation. Others were even decommissioned. It has been said that the most powerful nuclear industry in the world was no longer struggling with candle-light marchers and the sit-ins in nuclear plants, but with investors on Wall Street and the public service commissions. According to some eminent nuclear scientists, the accident also marked the beginning of a worldwide withdraw from investment in nuclear power production under the prevailing technological conditions.

The fire in one of the nuclear reactors in the Chernobyl power station, near Kiev, was by far the worst accident in the history of nuclear energy up until the present time. The reactor is now embedded in a sarcophage, an entombment of several thousand tons of sand, clay, lead, and boron, which had to be dropped from helicopters to extinguish the fire and prevent criticality in the demolished reactor core.

This accident occurred during shutdown of the reactor for routine maintenance, at which time an experiment was carried out and the reactor staff did not respect the operating regulations. The investigation of the accident points primarily to "gross human error": in order to perform the experiment, the operators deliberately violated the safety rules. The reactor became uncontrollable, causing a surge of power and heat that resulted in an explosion and the ejection of enormous amounts of radioactive material. The logbook revealed that it took less than 1 minute for the runaway reaction to destroy the reactor unit and initiate the chain of disastrous events.

The Chernobyl reactor was a graphite-moderated and water-cooled assembly using slightly enriched uranium oxide. None of these features, taken individually, is unique to Soviet design; they are also employed in various military reactors in the United States. There were, nevertheless, certain specific aspects. For example, there were more than 1600 separate pressure tubes through which water circulated around the fuel elements to remove heat. Steam was accumulated in huge drums before being conveyed to the turbine. The core of the reactor was very large, being about 12 meters in diameter and 7 meters high, and contained 2500 tons of graphite.

In the accident, inadequate cooling caused a surge of heat and fire; the latter consumed nearly 10 percent of the graphite. The fuel overheated, it fragmented into minute hot pieces, and these pieces very rapidly vaporized the water present in the cooling system. This very fast vapor production resulted in a shock wave which mechanical energy was probably equivalent to a few hundred kilograms of TNT. Its explosions lifted the roof off the reactor vault and catapulted redhot lumps of graphite and pieces of uranium into the surrounding buildings and spaces.

The Soviet authorities estimated that direct losses alone amounted to about 2000 million rubles, the equivalent of about 2800 million dollars. This figure presumably included the cleanup operation and the lost output for the plant itself, together with the outputs for farms and factories in the surrounding contaminated area of about 1000 square kilometers. Three months after the accident, the casualties appeared to involve 31 deaths, with a further 200 persons seriously affected by radiation sickness. All persons from the contaminated zone (135,000) were evacuated and many of them resettled. In the accident area, more than 500 settlements and about 60,000 buildings were submitted to thorough decontamination in the course of the following 7 months. Extensive ramparts were constructed to protect the rivers which supply the area with drinking water from contaminated underground streams.

The activity released in the accident amounted to about 50 million curies, i.e., 2×10^{18} becquerels, for nongaseous fission products, iodine and the noble gases argon and xenon. These values were based on measurements of fallout in the Soviet Union, and are considered by Soviet authorities to be correct within 50 percent.

A detailed report of the Soviet government to the International Atomic Energy Agency provided abundant information on the accident and its immediate consequences. The population of 24,000 persons at distances of 3 to 15 kilometers from the plant received or is expected to receive an average of 0.44 sievert (44 rems). This information makes possible an assessment of the cancer risk by taking into account the considerations outlined in Chapter 3, where it was noted that between 12.5 and 50 fatal cancers per millisievert are expected in a population of 1 million persons (i.e., an additional 12.5 to 50 fatal cancers per 1000 man-sieverts). With respect to Chernobyl, 24,000 men × 0.44 sievert = 10,560 man-sieverts, which means that between 132 and 528 fatal cancers could affect this population group as a result of the nuclear accident. Similarly, it is estimated that the population of 75 million people living in Byelorussia and Ukraine, who received 290,000 man-sieverts over their lifetimes (i.e. on the average 3.9 millisieverts per person per lifetime), will reveal 3600 to 14,500 fatal cancers due to the Chernobyl accident.

Because of the favorable wind direction, the population of the nearby town of Pripyat was exposed to a smaller extent of radioactive fallout and suffered less irradiations. This community is therefore not included in the consideration relevant to the Chernobyl neighborhood. Firemen and reactor operators received the greatest exposure during the accident. Among this group 20 were classified as most severely injured, and 17 died within 6 weeks. They received full-body doses ranging from 6 to 16 grays (600 to 1600 rems). Many of them suffered lethal skin burns from β-radiation.

According to Soviet plans the health of 135,000 people evacuated from the contaminated area will be monitored during the next 40 to 50 years. This action is universally considered as a most important undertaking in establishing whether there exists a radiation dose threshold for the appearance of biological effects.

Analyses of fallout throughout the European continent showed that for 200 million people in Eastern and Central Europe who were exposed to Chernobyl's radioactive cloud, the additional radiation dose would be about the same as that received by generations of humans during the early 1960s, when populations were exposed to global fallout from atmospheric nuclear tests. Nevertheless, the public at large was alarmed by the increased radioactivity of air, soil, and particularly of food. Confusion was caused when many governmental agencies released contradictory or controversal information and recommendations. There was an evident lack of communication between countries even inside the European community. In Switzerland, for example, people who watched German television were told that the milk was safe, whereas the Italian television warned against drinking the milk; at the same time, French viewers were told very little at all.

What is Safe Enough?

The lesson of the accidents in Windscale, Three Mile Island and Chernobyl is that the existing safety systems are not sufficient in the event of emergency. Even if the operators press all the wrong buttons, the reactor should be sufficiently robust and stable to leave enough time for the personnel to realize the mistake and correct it before damage is done. Alternatively, the reactor's construction must be inherently safe without requiring the intervention of man or machine. The three major accidents considered here show that the nuclear industry still has to learn how to make a reliable and safe nuclear power system.

But "what is safe enough"? One major accident in 1000 reactor-years is certainly a high rate of incidence in a technical civilization which already possesses several hundred reactors, and consequently an accident could be expected to occur every few years.

The reliability of a given system, such as a nuclear reactor, can be considered analytically in an attempt to foresee the accident and its consequences. The analysis is based on what is called "event trees" and "fault trees". An event tree identifies a sequence of events which can lead to failure. It grows from an initial event and seeks all possible outcomes. Next, it examines the sequence of all events that follow, including the information on safety features in the system. A fault tree inserts the probability of success or failure of the various individual components or systems on the event-tree path. By considering the overall system in this way it is possible to calculate a performance probability and the consequences of various potential accidents.

About 20 comprehensive reliability studies of the worst type of nuclear reactor accident, the core melt, were published by the mid 1980s. The results vary by as much as three orders of magnitude. According to an optimistic approach, one core fusion should be expected to occur in 1 million reactor-years; another estimate indicates only 500 reactor-years. The latter seems to be closer to reality, particularly if one takes into account not the total number of reactor-years, but the number relevant to the operation of a given type, for example, the PWR type or the type of reactor in Chernobyl.

The "realistic" result may nevertheless be fortuitous. The reason is that the studies of risks were based on the rather limited experience available in the middle of the present decade, and on the scanty information on accidents that the nuclear energy authorities throughout the world were willing to communicate. Better information than that available at present is certainly of paramount importance for any analytical evaluation of the reliability of nuclear reactors. Furthermore, other approaches to the evaluation of risks appear desirable. It has been suggested that for nuclear reactors, as for earthquakes, the risk evaluations should deal not only with the probability of catastrophic events, in the "low probability-high risk" area, but also with other types of accidents of various magnitudes and frequencies.

At the beginning of the 1990s an international nuclear event scale was in preparation (INES). It categorizes events from level zero, for no safety significance, to level 7, for accidents with widespread health and environmental consequences. The idea is that the INES also covers nonreactor nuclear installations, such as fuel fabrication and enrichment facilities, as well as facilities involved in reprocessing, transportation, and waste storage and disposal.

NUCLEAR EXPLOSIVES

With increasing concentration of fissile material, either ^{235}U or ^{239}Pu, criticality is easier to achieve since more neutrons become available to start and maintain the chain process. In a mass of pure fissile material, this process can be triggered by a single neutron from cosmic rays. The instantaneous burst of criticality ends rather as an expulsion of heat and radiation than an explosive release of nuclear energy.

The Fission Bomb

For a nuclear explosion to occur, the fission process must proceed long enough for the explosive energy to build up before the rapid expansion and separation of the critical mass interrupts the chain. To allow enough time for the full force of the explosion to develop, the exploding mass must be held together so that its components will not fly apart too rapidly. This is achieved in a device called the fission ("atomic") bomb.

The fission bomb consists of two or more subcritical lumps of fissile material, which are brought together to form the critical mass only when the bomb is activated. The assembly of subcritical lumps must be very rapid and achieved within a millionth of a second. One way of accomplishing this is to surround the lumps with a chemical explosive which, when ignited by an electrical discharge, drives the lumps inwards, and compresses the fissile material into the critical mass. A neutron source placed in the core of the device initiates the chain fission of ^{235}U or ^{239}Pu. To retain the exploding mass in a compact form as long as possible, a "tamper" of natural uranium is used. Its inertia holds the critical mass together for a microsecond which is sufficient for the explosion to develop (Figure 38).

The power of a fission device is limited because the total mass of fissile material must be restricted. In order to avoid spontaneous criticality in the lumps of fissile material which make up the bomb, each of them must contain a subcritical mass of fissile uranium. As a result the total mass of fissile material, and hence the power of a fission device, cannot exceed a certain value. This upper limit of destructive power for a fission bomb is equivalent to 500 kilotons of TNT, a common chemical explosive.

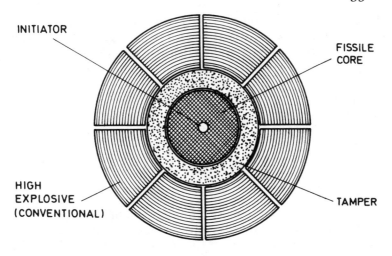

FIGURE 38. Diagram of a fission ("atomic") bomb.

The power of a nuclear explosive is expressed as its equivalent of TNT, in kilotons or megatons. This is a rough comparison, since the characteristics of a nuclear and a chemical explosion differ widely. The Hiroshima bomb was equivalent to 12.5 kilotons. The largest nuclear weapon, which was tested already a quarter of a century ago, corresponded to 50,000 kilotons, i.e., 50 megatons. It was based on another type of nuclear process known as fusion.

The Fusion Bomb

In the fusion process, light nuclides are merged to produce a heavier nucleus, whereby an enormous amount of energy is liberated. A nuclear fusion explosive has a particular advantage over the fission bomb in that no critical mass is required. Any amount of deuterium, for example, can sustain fusion; the larger the mass, the higher the amount of energy released during fusion.

However, the fusion bomb as a unit in itself cannot function. The nuclear fuel must be preheated to a temperature of tens of millions of degrees so that the repulsive forces between the nuclei can be overcome and the fusion process initiated. The fission bomb provides such temperatures, and is therefore used as a trigger in association with the fusion device, which is usually called a thermonuclear or "hydrogen" bomb.

The bomb (Figure 39) contains a material rich in "heavy hydrogen", i.e., deuterium, usually as a compound with lithium, and surrounds the trigger. Most fusion bombs also have a "tamper" of natural uranium as an outer layer. This absorbs energetic neutrons, which induce the fission of ^{238}U. Accordingly, a nuclear weapon of this type is in reality a fission-fusion-fission device.

In the absence of a uranium tamper the neutrons escape from the bomb as a deadly, penetrating radiation. This is the principle of the neutron bomb.

Effects

The immediate effects of a nuclear explosion are those of blast, heat and radiation. In addition, fission products appear as fallout. Energetic radiation in the form of γ-rays and neutrons represents about 5 percent of the total energy released by a fission bomb of a few kilotons, and up to 30 percent in the case of a neutron bomb.

The consequences of the use of nuclear explosives as weapons are considered in detail in Chapter 10.

Other Uses

Attempts have been made to use nuclear devices for nonmilitary purposes, as a more powerful substitute for chemical explosives. An example is the extraction of petroleum. The rock walls of a hydrocarbon reservoir often have low permeability, and rarely more than a quarter of the available oil can be extracted. The fractures caused in the reservoir by use of a nuclear explosive permit a freer flow of oil and gas to the wells. It has been found that this technique can increase the yield of raw petroleum by as much as 60 percent. However, the oil and gas are radioactive, and their delivery to the market requires an unambiguous assessment of the risks and benefits of production methods based on nuclear explosives.

Between 1965 and 1983, over 80 nuclear explosions, each of about 150 kilotons, were used in the Soviet Union alone in the prospection of oil and gas or in the excavation of reservoirs for storage.

For good reasons, public opinion is strongly against the "peaceful" uses of nuclear explosives. Each explosion contributes to the increase of global radioactivity and has harmful environmental consequences, in particular with respect to the biosphere. Moreover, such experiments can be — and most likely already have been — a screen for military purposes, such as the testing of nuclear warheads.

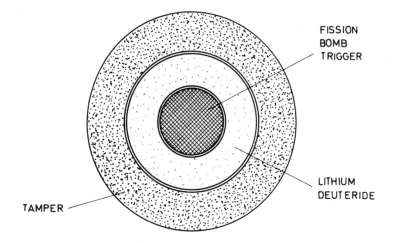

FISSION
BOMB
TRIGGER

LITHIUM
DEUTERIDE

TAMPER

FIGURE 39. Diagram of a fusion (thermonuclear) bomb.

RISKS AND BENEFITS

Nuclear power has been born and raised mainly in the framework of military programs to supply nuclear explosives and provide power for the propulsion of surface vessels and submarines. The generation of electricity was initiated as a by-product of the production of plutonium for military purposes.

After half a century of research and development in the field of nuclear power, our civilization is provided with some sixty thousand nuclear warheads, about six hundred nuclear-power submarines, cruisers, and aircraft carriers, and about four hundred nuclear power stations which generate some 12 per cent of the world's production of electricity. A consideration of the risks and benefits of nuclear power, at the beginning of the 1990s, cannot overlook these facts.

Explosives

The stockpile of nuclear weapons has long since reached a level sufficient to destroy our civilization. There is one nuclear warhead for every few tens of thousands of people on the Earth. Various analyses of the global impact of nuclear explosions on the biosphere show that there would be no such thing as a limited nuclear war. Under certain conditions, the extinction of the human species, and even of the whole biosphere, would not be unlikely (Chapter 10). Yet the gigantic military nuclear systems, with well-equipped laboratories and sophisticated production facilities, still are continuing their development beyond the view and knowledge of the public. Advocates of new nuclear weapons justify their existence as a necessary response to present or future threats in a world dominated by fear and misunderstanding. They claim that there is a benefit in nuclear weapons: their existence makes a world war, in particular, a nuclear one, obsolete.

This argument is not easy to accept even if the sense of responsibility of the present members of the "nuclear club" is recognized. The increasing number of nuclear reactors facilitates the access to fissile material, and knowledge and means of constructing nuclear weapons are within the reach of many countries outside the nuclear club.

Serious concerns also remain about the fate of the remaining 27,000 nuclear weapons in the former Soviet Union. There is apprehension that the nuclear weapons are not being stored securely. There also exists an uncertainty concerning the transfer of nuclear expertise and weapon-grade uranium or plutonium to countries outside the former Soviet Union which could help them acquire nuclear weapons.

The proliferation of nuclear weapons increases the risks of accidents, nuclear terrorism, and the inadvertant beginning of a nuclear war. For these reasons, nuclear weapons present an enormous, ever-increasing danger. Even with a complete ban on their production, testing, and use, and with a strict international control of fissile material and military installations, a considerable risk subsists.

The latter is certainly also associated with the so-called peaceful use of nuclear explosives. The excavation of underground storage spaces for gas or oil with the help of nuclear explosives is usually represented as an operation involving a low degree of contamination because of high dilution of radioactive material with large amounts of fossile fuel. In reality, such an interpretation is based on a fairly unlikely condition, namely, that the mixing of the oil or gas with fission products or other radioactive substances with underground materials is perfect. It also neglects another important fact. Diluted or not, large amounts of radioactivity released by the explosion and the subsequent use of fuel reach the surface and become dispersed in the environment. The resulting impact is qualitatively the same as that of any other fallout whether due to nuclear tests in the atmosphere or the accident at Chernobyl.

Electricity

When nuclear power plants first went into operation, some enthusiasts believed that electrical power would be so cheap that it would not be necessary to install meters for private consumers. It was also hoped that nuclear energy would provide the key to a bright future by providing abundant amounts of energy for a long time. The truth is that nuclear energy has indeed contributed to the global production of electricity, but it lags far behind the expectations of competent bodies such as the International Atomic Energy Agency of the United Nations. Of the various reasons for this, one is certainly that the basic philosophy concerning nuclear power plants has changed throughout the years. From "as cheap as possible" the motto has become as "safe as possible". More recently, the latter imperative has changed to as" safe enough". But how can these installations be made "safe enough"? The fear of accidents such as melting in the reactor core, particularly as in the incident at Chernobyl, is alarming not only public opinion, but also the authorities who license the nuclear plants. What is the acceptable risk for a major accident in a world in which about 400 nuclear power plants are presently operating, or in which a thousand may be in operation in a not-too-distant future? The answers depend on further technical progress, such as an increase in the reliability of reactor materials in minimizing radiation damage, or improvement in reactor technology which will prevent failures in cooling systems. Above all, there is a need for an inherently safe nuclear reactor system which is less dependent on the soundness of human judgment or the quality of technical components (Chapter 10).

Further Reading

Eisenbud, M., *Environmental Radioactivity from Natural, Industrial, and Military Sources*, 3rd ed, Academic Press, New York, 1987. [Emphasizes the role of artificial radioactivity; considers nuclear reactors, fuel reprocessing, and weapons. Gives details relevant to environmental dissemination of radioactive wastes, and information on nuclear waste managements, fallout, and nuclear accidents. Biological effects of radiation, radiation standards, safety, and risks are clearly presented. Valuable both for the professional and the general reader.]

Nuclear Power, the Environment and Man, International Atomic Energy Agency, Vienna, 1982. [A concise, clearly written survey of various aspects of nuclear power, fuel cycles, nuclear wastes, nuclear safety, and radiation protection.]

Patterson, W.C., *Nuclear Power*, Penguin Books, Hardmondsworth, England, 1985. [A popular presentation of basic principles and various technical aspects of nuclear reactors, the uranium fuel cycle, and radioactive wastes. The political, military, and economic implications of nuclear energy are critically analyzed in detail.]

Le Dossier Electronucléaire, Editions du Seuil, Paris, 1980 (in French). [Presents technical data on nuclear power plants and nuclear fuel. Provides valuable information on harmful effects of radiation and risks due to the uranium fuel cycle, describes accidents in nuclear power plants, and discusses environmental consequences of nuclear power production. Gives a detailed presentation of regulations and legal aspects of work in the nuclear industry (manufacture of reactor fuel and operation of electric power stations). Of great value is the survey of economic and industrial data relevant to nuclear energy production in France and elsewhere. A book that should be consulted because of the wealth of data collected with great care, experience, and honesty.]

Guéron J., *Les Matériaux Nucléaires*, Presse Universitaire de France, Paris, 1977; *L'Energie Nucléaire*, Presse Universitaire de France, Paris, 1982 (in French). [These two small books, written clearly by a highly experienced author, provide an excellent general picture of nuclear energy and a concise, popular presentation of numerous scientific and technical details.]

Wymer, R.G., and Vondra, B.L., Eds., *Light Water Reactor Nuclear Fuel Cycle*, CRC Press, Boca Raton, Florida, 1981. [A thorough consideration of various elements of the light water reactor fuel cycle, with emphasis on the chemistry of the fuel cycle operation.]

Throughout this book we have followed radiation, radioactivity and nuclear energy on the cosmological scale, from primordial radiation, the early appearance of elementary particles and earliest nucleosyntheses, to interstellar and interplanetary events on our early planet and the present-day Earth. This chapter deals with some perspectives.

How far into the future we can go depends on cosmic nuclear events: the end will come most likely about 5000 million years from now, when the nuclear processes that release energy in the Sun cease because of depletion of fuel in our star's core. Cosmic ray showers from a supernova explosion present a potential but rather remote danger for our planet and its biosphere.

How far into the future it is meaningful to go depends on various terrestrial nuclear events. One of particular importance is the use of the huge arsenal of nuclear weapons. Any widespread use would entail both severe radiation-induced damage and climatological changes with disastrous consequences for civilization, if not for the very existence of the human species and for the basic conditions which support life on our planet.

Fortunately, not all perspectives are necessarily somber. Nuclear power may provide a possible solution for the energy supply of an increasing world population, at least until satisfactory advances in environmentally cleaner energy sources are made.

Energy production based on nuclear fission is a reality. Its near future depends strongly on success in developing inherently safe nuclear reactors.

Controlled nuclear fusion could conceivably provide reasonably "clean" energy from a practically unlimited fuel supply. However, despite present spectacular achievements in laboratories, it would be a formidable task to produce a device which would maintain stellar temperatures long enough to produce electricity, especially on an industrial scale. For a satisfactory solution in a not too distant future, international cooperation is essential. Encouraging prospects are already in view.

Life on the Earth has developed since its beginning in the presence of ionizing radiation, but the nuclear age has brought the risks of accidental irradiations of large populations and dose levels which could be far higher than natural ones. Various coordinated international efforts to reduce the risks have been made following the Chernobyl accident, and some of the important goals include a global radiation monitoring system and international control of the radioactivity released into the environment.

10

Perspectives

NUCLEAR EVENTS BEYOND EARTH

Celestial surroundings are undergoing change while our planet travels on its long-continuing course around the Sun in the Solar System, with the latter through the Milky Way galaxy and, in turn with this, through intergalactic space. Two types of nuclear events outside the Earth are of particular interest to us: the possible occurrence of a supernova explosion in the proximity of the Solar System, and the nuclear fusion process — especially with respect to its duration — in the Sun. We can only surmise on how close to us supernovas may explode, or how intense the subsequent cosmic showers would be. However, we have fairly good understanding of the fate of the fusion reactor in the core of the Sun. The occurrence of other nuclear events and terminal processes can also be examined in the light of a standard model of the universe.

When the Fusion Reactor in the Sun Ceases

We have seen how a star core operates as a fusion reactor, burning hydrogen to produce helium and releasing enormous amounts of energy (Chapter 5). The amount of hydrogen fuel that can be burnt in a star is limited; in the Sun's core it is 10 percent. Until now, the Sun has consumed only 5 percent, which means that there is still enough hydrogen fuel for the next 5000 million years.

The Sun's fusion reactor operates by changing its fuel. Helium produced by the fusion of hydrogen has accumulated ever since the birth of the Sun, whose gravitational contraction further heats the core. The zone of hydrogen fusion slowly migrates outwards like an expanding shell. When it reaches a point where the temperature is less than 10 million degrees, the fusion process ceases. In the meantime, the internal gravitational field forces the helium-rich core to contract, thereby increasing the pressure and raising the temperature to 100 million degrees. This initiates the fusion of helium nuclei, causing a further production of nuclear energy and the syntheses of carbon and oxygen nuclei. The external layers of the star will expand and cool, and our Sun will then undergo a major change becoming a red giant star. In its expansion, it will envelop and devour the inner members of the Solar System, that is, Mercury, Venus, and also the Earth. After the helium is almost consumed, the interior region will continue its postponed collapse, and the Sun will eventually end as a white dwarf. The latter bodies represent a class of small, extremely brilliant, and highly dense stars with a high surface temperature (hence the color).

Cosmic Showers Due to a Supernova Explosion

Supernova explosions are expected to occur in any galaxy once every 20 to 50 years. The question of interest is how distant these events are, and how significant the danger of cosmic ray showers is for the life on our planet.

The present dose rate on the Earth's surface due to the normal cosmic ray flux is 0.3 milligray per year. If a supernova explosion occurred at a distance of 30 light-years, the dose might increase 1000-fold. Annual doses at this level on a geological time scale would have a significant influence on life, but would not have fatal consequences. The latter conditions would require cosmic showers corresponding to an annual dose of 3 grays over a longer period of time. This 10,000-fold increase in the dose rate would expectedly be produced by a supernova explosion at a distance of 10 to 15 light-years from the Earth. The recent and "nearby" supernova explosion recorded in 1987 has occurred at a distance of 170,000 light-years (Chapter 5).

In our cosmic vicinity within 10 to 30 light-years, there are no stars whose masses or behaviors point to supernova. The danger is elsewhere, in the spiral arms and regions within a few thousand light-years of the galactic center, where the abundance of massive stars and, hence, the probability of supernova explosions are high. Our Sun passes through the spiral arms of the galaxy only once every 100 million years, and takes 10 million years to complete the passage. Accordingly, our "planetary spacecraft" seems to be marginally safe for at least 100 million years to come.

It is possible that the Earth may have experienced in the past the increased irradiation due to supernova explosions. Many scientists have argued that such an event was responsible for the extermination of numerous forms of life some 60 million years ago, when the dinosaurs disappeared. In addition to the high intensity of cosmic rays, radiations from decaying radionuclides could have contributed significantly to the biological damage. The radionuclides were produced in the bombardment of terrestrial target nuclei by the neutrons of cosmic showers, and became incorporated into the biosphere. As localized sources of ionizing radiations, they effectively produced free radicals which could have been responsible for a multitude of chemical and biological changes.

The extinction of the dinosaurs is considered to be only one case in a series of catastrophic events which apparently occurred periodically in the past. Other explanations have been proposed, such as a collision of the Earth with a comet or an asteroid, with subsequent climatologic changes similar to those predicted in case of widespread use of nuclear weapons.

Toward the End of the Universe

We live in an expanding universe whose fate, according to the standard model, depends on whether the density of cosmic matter is lower or higher than a certain critical value.

Estimates based on the weight of all luminous (i.e., visible) celestial objects indicate a very low density, corresponding to less than 10 percent of the critical value. This is not substantially increased by further taking into account the mass of hydrogen, either atomic or ionized, which is the predominant chemical element in the universe. Even if it is assumed that there exist enormous amounts of nonluminescent matter in the galaxies in the form of cold rocks, extinguished stars, and black holes, the critical value is not attained. Additional contributions of dark matter, neutrinos, or even of hypothetical axions (Chapter 5) still seem to be far from providing a satisfactory explanation.

If the universe is dense enough, gravitational attraction should slow down its expansion and eventually reverse the process. The contraction of such a universe, consisting largely of a mixture of dead stars and black holes, would release unimaginable amounts of heat. This would result in a "soup" of particles, and the universe would tumble into a "big squeeze". Thousands of millions of years from now, the process should produce a new "cosmic egg", that will explode in another Big Bang, and the process will start all over again. If, indeed, the density of the universe is low, as it appears to be, we may live in a cosmos that will expand forever. In this case the radioactivity of the proton, if experimentally confirmed (Chapter 2), will play a decisive role.

It is considered that the proton may decay with a half-life of 10^{32} years. Positrons, neutrinos and photons should be produced in the decay. The universe, which has never ceased to expand, will become a very rarefied gas of leptons (Table 4, Chapter 2), and photons.

NUCLEAR WAR — A MAJOR DANGER

The Arms Race

The public at large became aware of the nuclear arms race in the 1960s when the possibility of extermination on a global scale already existed. At the time, there were about 2500 bombers and missiles and three times as many nuclear weapons. The explosive power was equivalent to 8000 megatons of trinitrotoluene (TNT), and the nuclear arsenal was shared by the United States and the Soviet Union, with Great Britain and France only beginning the race.

A survey conducted some 20 years later revealed 50,000 warheads in strategic and tactical nuclear weapons, with an explosive power equivalent to from 12,000 to 15,000 megatons of TNT, or about 1 million Hiroshima bombs. Thus, at the beginning of the present decade, one Hiroshima bomb was available for every 5,000 inhabitants on the Earth. After 3 decades of the nuclear arms race, each inhabitant of our planet had his equivalent of explosive corresponding to 3 tons of TNT.

At the end of this last decade, the overkill potential was distributed among the arsenals of five countries: the Soviet Union, the United States, France, Great Britain, and China. The first two alone disposed of 97 percent of the nuclear warheads, equipped with a large variety of sophisticated carriers. According to data released in 1985 for the inventories of strategic weapons, Great Britain possesses 64 and France 212 nuclear warheads. At the same time, estimates of the Chinese arsenal indicated 580 warheads in both strategic and tactical nuclear weapons. The term strategic weapons signifies those which are delivered by long-range vehicles such as bombers and intercontinental missiles, or by submarine-launched ballistic missiles. Tactical weapons are used on the battlefield, e.g., in tanks or aircraft carriers.

The beginning of the 1990s has seen a drastic change in the behavior of the two nuclear superpowers, from arms race and conflict situations to cooperation (Chapter 1). Serious concerns, nevertheless, remain because of political and social instability in the former Soviet Union, which still owns almost half of the global nuclear arsenal and capacity for its production.

When speaking of nuclear weapons, one usually refers to the arsenals of the two superpowers. However, the role of strategic nuclear forces of other countries in the world theater is much more significant than would appear on the basis of their apparently low contribution to the share of nuclear arsenals. A numerical comparison evades the essential point. It is certain that the European or Chinese forces, as they stand now, could inflict serious damage to the country they consider as an enemy. Also, since they are not at present bound by any form of arms treaty, their very existence complicates negotiations between the superpowers on nuclear arms control and limitation. Further proliferation of nuclear weapons, which is already within the reach of several countries in Europe and Asia, would only serve to increase the risk of a global nuclear conflict. Even on an initially limited basis, a nuclear confrontation could drag the world into a full-scale nuclear war.

Nuclear weapons are dreadful not only because of their number. There is no point on the globe that cannot be reached with the available means of delivery, which comprises missiles, bombers and submarines. The precision that can be attained lies within a deviation of 300 meters, which is sufficient to reach any target, whether it be a city or a silo. The energy released by a single bomb of several megatons is enough to incapacitate the largest city in the world. After the tests with bombs of a few dozen megatons some 20 years ago (U.S. 15 megatons; U.S.S.R. 57 megatons), no efforts were made to design more powerful devices.

It has been assumed for years that the most devastating consequence of a major nuclear conflict would be the large number of human casualties in the principal target areas (e.g., the Northern Hemisphere), probably entailing a complete collapse of social and economic structures. It was believed that nations not engaged in the conflict, and thus the majority of the human population, would not be endangered. The global consequences would involve only indirect effects such as a delayed radioactive fallout, minor climatic changes because of dust and smoke injected into the atmosphere, or a depletion of ozone in the upper atmosphere.

The prevailing opinion at present is that no one would be left unaffected, and that a unilateral, single nuclear blow would be lethal even for the aggressor. Furthermore, it is believed that the beginning of a nuclear conflict is not necessarily limited to one of the superpowers. It could start elsewhere, even accidentally, and continue with little chance of remaining localized. Many comprehensive studies of the aspects and consequences of a nuclear war point to a holocaust as the most likely result.

A Megaton Nuclear Weapon

As an example, we consider a typical strategic warhead, such as that used in intercontinental ballistic missiles. The heat generated by detonation vaporizes the weapon almost instantaneously and the nuclear process stops. Most of the nuclei produced by the fission of uranium are highly excited and return to stable states by the emission of energetic X- and γ-rays. The radiations heat the surrounding air, forming a shock wave that heats additional layers of air. The result is a luminous fireball which rises to an altitude of approximately 20 kilometers in 2.5 minutes.

The explosion of a 1-megaton bomb can excavate a crater up to 400 meters in diameter and more than 100 meters deep. Large amounts of soil and debris are lifted with the updrafts generated by the rising fireball. Heat, blast, and penetrating radiation destroy everything within a radius of 3 kilometers; the center of this circle is the so-called ground zero, the point on the ground (or above it for an explosion in the atmosphere) at which the nuclear weapon detonated. Within this area of about 30 square kilometers, the overpressure due to the explosion is at least 1 kilogram per square centimeter above the normal atmospheric pressure; the temperatures attain thousands of degrees and the radiation doses thousands of grays. Most objects in the area are first pulverized and then melted or vaporized. Any form of living matter is reduced to smoke and ashes and simply disappears.

At greater distances, the effects of radiation are less pronounced, although those of blast and heat are still important and are felt up to 4 and 7 kilometers, respectively, from ground zero. In this area the fatalities are mainly due to collapsing buildings, flying debris, and fires; the survivors, corresponding to about half of the population in the area, will suffer numerous injuries. Their treatment would be problematical in the absence of available medical care and the situation would be aggravated by a lack of potable water, and energy supplies such as electricity and gas. In addition to these consequences of blast damage and fires, there would be long-continuing irradiation and radioactive contamination.

Radioactive nuclides created by the explosion condense on dust particles in the air as the fireball cools. These return later to the ground as radioactive fallout. In fact, the latter forms and drops almost immediately after the explosion, since 70 percent of the fallout consists of larger particles. They are deposited within the same day, but the location will depend on wind conditions and may be at considerable distances from ground zero. The radiation level of the contaminating fallout can greatly exceed the lethal dose, thus killing many persons who survived the blast wave and the thermal pulse. The total radioactivity released by the explosion is enormous and may initially amount to more than 100,000 million curies.

A single-megaton nuclear weapon can devastate a substantial part of an industralized, densely populated country. This is especially true if the target includes a nuclear reactor, for example, in a nuclear power plant operating at 1000 megawatts of electricity. If the reactor fuel is completely evaporated by the explosion, its radioactivity will combine with that released by the bomb, rising with the fireball and returning to the ground with the fallout.

Although the origin of radioactivity in both cases is uranium fission the relative contributions in this example differ considerably. The shorter-lived radionuclides decay during the reactor operation, prior to the explosion, and the debris from the reactor fuel thus contributes significantly to the long-lasting radioactive contamination. Fallout radioactivity due to the bomb explosion, on the other hand, is initially dominated by the short-lived fission products (Figure 40).

The total dose rate would be lethal within an area of more than 1250 square kilometers. For a high population density as in Europe, several hundred thousand people would be killed. Furthermore, a surface of 22,000 square kilometers would require evacuation, remaining uninhabitable for at least a year, assuming that a total annual dose of 0.1 gray is an unacceptable risk.

Analysis of a hypothetical nuclear attack provides additional information. Let us suppose that a 1000-megawatt-of-electricity nuclear power plant is struck near Stuttgart, in the Federal Republic of Germany. With a prevailing southeast wind of 24 kilometers per hour and after 1 month, the zone of harmful radiation with a dose of 0.1 gray per year would presumably extend over Frankfurt, Essen, Amsterdam, Rotterdam, and Great Britain. One year after the attack, the zone still would include many of the industrial parts of West Germany and Holland.

This horrible scenario by no means corresponds to the most pessimistic prediction, since the level of radioactivity involved could easily be several times higher. A multiple nuclear warhead could be directed simultaneously toward several nuclear reactors;

this is not a particularly difficult task, since a nuclear power plant site usually includes more than one reactor. The reactors are often constructed in pairs, and the four at the Chernobyl power station, only several hundred meters apart, are by no means an exception. Moreover, the spent fuel elements, which are frequently located in water pools in the vicinity of the reactors, may contain more and longer-lived radioelements than the reactors themselves.

In localities of dense population and intensively exploited land, the power reactors are inevitably close to military installations. Even when aimed at a military target, nuclear warheads could inadvertently destroy a nearby reactor, or the pool of spent fuel.

A 10-Kiloton Nuclear Weapon

This can be taken as an example of a tactical nuclear weapon. Upon explosion, the ranges of direct thermal, blast, and radiation effects are confined to a zone approximately 1 kilometer in radius. The overpressure would be 0.5 kilograms per square centimeter and the blast strong enough to cause severe damage to buildings at distances up to 1 kilometer. The instantaneous radiation doses would be 80 grays at 690 meters from ground zero, 1.50 grays at 1300 meters, and 0.30 gray at 1500 meters.

These effects may seem relatively mild when compared to those of a 1-megaton nuclear weapon. Yet, a similar weapon of about 12.5 kilotons, launched on Hiroshima 4 decades ago, killed 130,000 persons, and injured numerous others among an overall population of 340,000. Following the explosion, which occurred 600 meters above the city, the central urban area was transformed into a flat, rubble-strewn plain from which only a few ruins of sturdier buildings protruded; 68 percent of the constructions were either completely destroyed or damaged beyond repair. Half an hour after the blast, which occurred at 8 a.m., the fires kindled by the thermal pulse and the collapse of buildings began to coalesce into a wildfire which lasted 6 hours. A "black rain" carrying radioactive fallout fell on parts of the city. During the forenoon, a violent whirlwind further devastated the city. The weather was fair at the moment of bombing, and wind and rain were the result of peculiar local meterological conditions generated by the explosion.

Massive Use: Direct Effects

Blast, thermal pulse and prompt radiation are the direct effects of nuclear weapons. They are well understood not only from

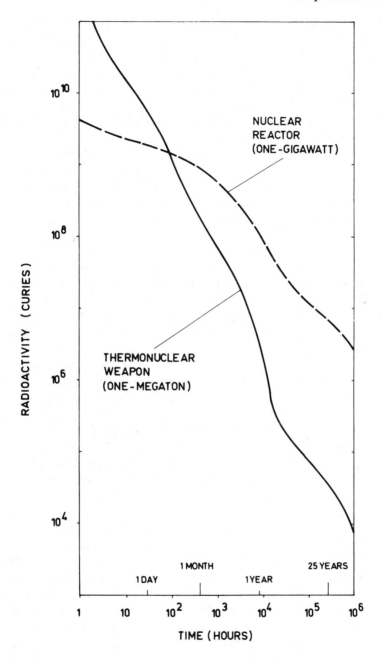

FIGURE 40. Decay of radioactivity released by the detonation of a 1-megaton nuclear weapon and the core of a 1-gigawatt nuclear reactor.

the two detonations on the Japanese mainland at the end of World War II, but also from numerous atmospheric and underground tests on military polygons. In the period from 1945 to the end of 1985, a total of 1570 test explosions had been conducted. Numbers for various countries are United States, 805; Soviet Union, 562; France, 134; Great Britain, 39; China, 29; and India, 1. Their total explosive yield was equivalent to 750,000 kilotons of TNT. On these grounds of experience, the consequences of a widespread use of nuclear weapons can be anticipated. It is worth noting that the price paid for this experience was the release of more than 7.50 $\times 10^{13}$ curies of short- and long-lived radioactive elements into the environment. To get an idea of what this represents, we may compare this figure with the total activity of radioelements on the present-day Earth (Figure 28, Chapter 8).

The acquired experience clearly shows that in a total war fought with nuclear weapons there will be no winner, regardless who strikes first. Accordingly, the military lobby now envisages the limited nuclear war, which is scarcely a less dangerous concept. Even if a meaningful definition of a limited use of nuclear weapons could be given, its practical realization would still remain meaningless for a very simple reason: there is no "backyard" which could serve as a theater for the conflict, and in which the radioactivity and radiation could be confined. The Chernobyl accident, which was a minor event in comparison with even a single nuclear bomb explosion, is a good illustration of how radioactivity can spread beyond national borders.

The military establishments have worked on various "scenarios" of a nuclear war and on "options" for nuclear attack or defense. The conclusions of such analyses are more than frightening. For example, 18 warheads hitting New York, with its population of 16.3 million, would cause the death of 12.9 million inhabitants and injuries to another 3 million. To annihilate all urban areas in the United States, with a population of over 10,000 inhabitants, among a total national population of 188 million, only 1000 nuclear weapons would be necessary. The number of dead and injured together would be 169.5 million, of which the mortality figure would be 140.5 million. Here, as in all other figures relevant to nuclear weapon casualities, the number of deaths corresponds to persons killed instantly or who die within several weeks from radiation and other injuries. Several, often ten, nuclear warheads are carried by a single strategic missile, for example, of the intercontinental ballistic type.

A dozen megaton nuclear weapons would be sufficient to incapacitate most European countries. In this densely populated part of the world the immediate death toll would amount to 120 million persons.

Other Consequences

Fire and smoke have particular consequences in nuclear explosions.

Thermal pulses from widespread nuclear explosions would ignite large amounts of materials over wide areas, starting fires in forests and cities. Burning asphalt, oil, and coal in the cities would contribute to the total amount of smoke and soot more than burning wood. Part of the smoke sent into the atmosphere could quickly return in the form of "black rain". However, considerable amounts would remain in the atmosphere, absorb the Sun's light and warmth, and subsequently cause a drop in temperature.

Natural temperature variations are regulated by very complex processes in the atmosphere, and information on observed trends is best provided by computer simulation. Powerful computers and sophisticated programs for modeling are used, since numerous parameters have to be taken into account. Simulation of conditions following nuclear conflict has been carried out using a three-dimensional model, in which the smoke is free to move in any direction, and is warmed by the absorption of sunlight and washed down by rain.

The critical choice of input data is very important for computer simulation. Here, these data concern the amount of smoke, the number, type, and explosive power of the nuclear weapons considered, the number and kind of targets chosen, the height above the ground of each detonation, the season of the year, and the part of the globe where the event occurs.

The best choice of data is not a simple matter, since many of the parameters are still not well known. There are elements of doubt with respect to the characteristics of large fires, the heights to which the smoke will rise, the optical properties of smoke particles of different sizes, and mechanisms by which these particles are rained out. Consequently, there may be significant uncertainties in the calculations and the predictions can range from what is well known as the "nuclear winter" to the "nuclear fall". The latter term has recently been proposed to denote milder consequences. Nonetheless, regardless of the uncertainties in the input data and the nomenclature used to label the outcome, all models suggest that fires following a nuclear war would lead to a significant decrease in the intensity of light reaching the Earth's surface, thus causing a marked drop in temperature. The global **impact on the biosphere would be disastrous.**

No one can know in advance how a nuclear war will be conducted. Computer modeling is usually based on the assumption that a substantial fraction of the arsenal of major powers will be used, corresponding to a total of 100 to 10,000 megatons. We shall consider here results based on the use of 6500 megatons. In this scenario, 25,000 warheads of both strategic and tactical type are launched on a summer day in the Northern Hemisphere, toward a thousand of the largest cities and military installations in the NATO and Warsaw pact countries. After a rapid fallout of 50 percent of the pulverized debris, the amount of rising smoke is estimated at 30 to 150 million tons. Of particular importance for the light absorption is the presence of 30 million tons of amorphous elemental carbon. If spread out over the Northern Hemisphere, this could reduce the insolation at the level of the ground by 90 percent and cause a drop in temperature of between 15 and 35 degrees celsius within a few weeks following the attack. This prediction is plausible for the interior of the continent, whereas coastal areas and islands might experience a temperature decrease of only several degrees owing to the warming effect of the maritime region. The long-term consequences might last for years and could eventually be felt in the Southern Hemisphere as well, even if the attacks were limited to the northern part of the globe.

The effects on temperature would be half as great in the case of a "winter scenario". Nevertheless, even then the consequences would remain disastrous. A decrease of only 3 to 5 degrees celsius at the beginning of the growing season would be sufficient to destroy harvests in North America and the Soviet Union; the rice crop in Asia would fail if at any time during the growing season the temperature falls below 15 degrees celsius. A less certain, but not improbable consequence would be disruption of the world precipitation system. Alterations in the summer monsoons in Asia and Africa could seriously contribute to crop failures and loss of livestock. Various estimates suggest that in nations engaged in conflict a few hundred million persons might die from direct effects, but mass starvation on a global scale could kill several times this many.

The greatest uncertainty in the model lies in the data taken for the initial amount of smoke. A baseline, or average value, is set at 180 million tons, but plausible amounts range anywhere from 20 million tons — which would have negligible climatic effects — to 650 million tons, which might plunge not only the Northern Hemisphere, but the whole globe into a deep freeze.

The fires would also produce various toxic substances and the chemical air pollutants would include oxides of carbon and sulfur, hydrocarbons, hydrochloric acid, and asbestos. Especially noxious for human health would be the nitrogen oxides, which are abundantly formed at the high temperatures of nuclear fireballs. These oxides would initiate chemical processes in the stratosphere and deplete the ozone layer. Within a month, the ozone concentration could fall to the extent of 10 to 30 percent. It would take several years to replenish it; in the meantime, the increased intensity of ultraviolet light at the Earth's surface would cause damage to the biosphere.

The formation and accumulation of radioactive fallout would be another deleterious process in the atmosphere. Apart from the prompt fallout, which is generated directly by exploding bombs, there would be the long-term fallout due to particles carried far out into the stratosphere. In the scene of nuclear war considered here, roughly 30 percent of the land in northern midlatitudes could receive doses higher than 2.50 grays, and as much as 50 percent could be exposed to about 1.00 gray. The consequences would be very harmful for life on Earth in general, and for the human species in particular.

ENERGY AND SOCIAL EVOLUTION

A Contemporary Outlook

The annual global consumption of energy during the late 1980s was about 400 exajoules. Only 100 exajoules is consumed outside the industrial countries, which represent more than 70 percent of the world population. One exajoule (EJ) is 10^{18} joules and corresponds to 2.8×10^{11} kilowatt-hours. To produce 1 exajoule, 31 power stations rated at 1000 megawatts must operate for 1 year and consume 23 million tons of oil or 35 million tons of coal.

This difference in energy consumption reflects a striking degree of social inequality. In some parts of the world the stocks of food accumulate on oversaturated markets, whereas elsewhere famine causes the deaths of millions. The living conditions differ widely, and the average life expectancy, which is over 70 years of age in developed countries, is barely 30 in regions where food, energy, and medical supplies are insufficient.

Human progress is closely associated with development in energy production. Several hundred thousand years ago, the main energy source available to primitive man was the food he ate. The hunter in 100,000 BC had additional energy and food by burning wood for heating and cooking. The earliest farmers in 5000 BC were the first to use domestic animals as a further energy source. The farmer of 1400 AD began to exploit the power of water and wind, and the industrial man of 1875 AD started to use coal and steam. During the last hundred years, the energy consumption has increased almost 20-fold, the average life span has doubled, and the world population has tripled. By the end of this century the global population will expectedly rise to about 6000 million. In order to meet requirements and maintain development, the annual energy production will necessarily exceed the present 400 exajoules.

In principle, there should be no problem in the foreseeable future, since the Sun alone provides us with more than enough energy. In addition to solar radiation, this energy is available in indirect forms, from the wind and the sea, and from trees and other plants which convert the energy of sun rays into carbohydrates and thereby contribute to the storage of energy as biomass.

In practice, a suitable fuel is required. Before the beginning of the last century, wood was the predominant source of domestic energy. This was gradually replaced by fossil fuel, first coal and then oil and natural gas. In recent times nuclear fuel gained rapidly in importance until public opinion became alarmed by accidents involving nuclear wastes and the operation of nuclear stations.

Our energy reserves must be taken into account in all considerations of future energy supplies. The values given for the reserves of renewable sources and for fossil fuels concern available deposits, which are reliably known and can be economically exploited by established techniques. Availability is only one of the factors, and others which must be considered are the impact on the environment and our health, and the efficiency of conversion into mechanical work or other suitable forms of energy. The time needed to pass from discovery and the scale of laboratory experiments to commercial production is also of importance to any novel technique.

Renewable Sources

An analysis of various forms of renewable sources of energy (Figure 41) shows that each has its advantages and limitations. The impressive amount of 420 exajoules quoted for the biomass assumes that virtually all productive land on Earth be cultivated and that all crops be converted into some form of energy. This is evidently far from reality. The biomass is considered useful for moderate consumers such as in the rural communities of developing countries and at present it is not promising for large-scale use. Depending on the climate and the type of plants grown, several thousand square kilometers of land are needed to produce fuel for a 1000-megawatt-of-electricity power station. In order to avoid excessive consumption of water and fertilizer, new solutions must be found in selecting and using the crop as an energy source.

Water is certainly the energy producer of choice, because it is nonpolluting and does not consume atmospheric oxygen. Nevertheless, only 16 exajoules of the potentially available 126 exajoules

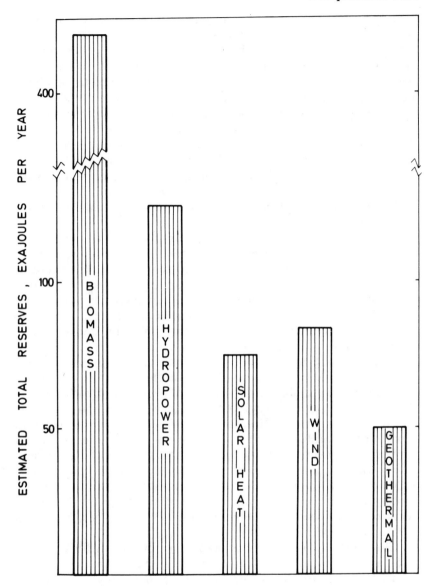

FIGURE 41. Reserves of renewable forms of energy.

are globally produced by hydroelectric power. For this kind of plant, geological constraints may be determinant, and the risks of dam failures must be taken into account.

Direct solar heat provides 75 exajoules per year, and this energy is spread over the globe at a moderate level. With presently existing techniques, this source is convenient for heating water and family homes, but is unsuited for large-scale production of electricity.

In the absence of our atmosphere, incident solar radiation would theoretically provide an energy of 1400 joules per second per square meter on the Earth's surface. Under natural conditions the amount of energy received is much smaller. With the aid of presently available solar cells, sunlight is nonetheless an effective means for the direct conversion of radiant energy to electricity on a limited scale. Depending on efficiency, and 20 percent is at present considered to be a maximum, 30 to 50 square kilometers of solar cells would however be needed to replace a single power station of 1000 megawatts of electricity. Understandably, the supply of energy in this way would be intermittent and vary according to weather, the time of day, and the season. Sunnier areas, such as those in Africa, could provide higher production rates and better economy.

Large-scale applications will depend on the future realization of orbital solar power plants, and on their means of conveyance to the Earth. Various barriers of technological, economic, and judicial nature have to be overcome before space solar technology can be adapted to large-scale terrestrial use. It will require some time, probably several decades, but basic solutions are already in view. Meanwhile, terrestrial use of solar energy will be limited to a short list of favorable sites and specialized uses.

The exploitation of wind and geothermal energy in the mid-1980s amounted to less than 1/1000 of the available reserves. A major disadvantage in the generation of electricity by wind is the lack of a means of energy storage and a stable supply to the consumer. Large areas of free land would be needed for generators capable of providing power at output levels above 1 megawatt. On the basis of present technology, a typical 30 meters high turbine with blades of about 30 meters in diameter, would have an output of 300 to 500 kilowatts. It has been estimated that a "wind-park" should have 25 turbines to generate sufficient electricity for the needs of 15,000 people in 1990s in Great Britain. The park would occupy 3 to 4 square kilometers.

The interior of our planet stores enormous quantities of heat, but as yet little effort is made for its extraction, except for occasional tapping of geothermal energy from geysers and hot springs. The water which arrives at the surface has usually a relatively low temperature, a significant salt content, and releases radon. The operation of injecting water into the deep-lying hot rock layers and retrieving it at the surface as a steam still remains a major challenge.

Fossil Fuels

Oil and natural gas can be used directly and are easy to store, transport, and distribute. Disadvantages include fire hazards and risks of transport accidents; oil loss and seepage are a common source of pollution. Their combustion consumes oxygen, and releases oxides of carbon and, in the case of oil, sulfur and suspended particles as well. Here, again, serious environmental pollution may occur since the exhaust gases are not usually purified.

The reserves of oil and gas (Figure 42) will be expectedly exhausted in the course of the next century. The remaining

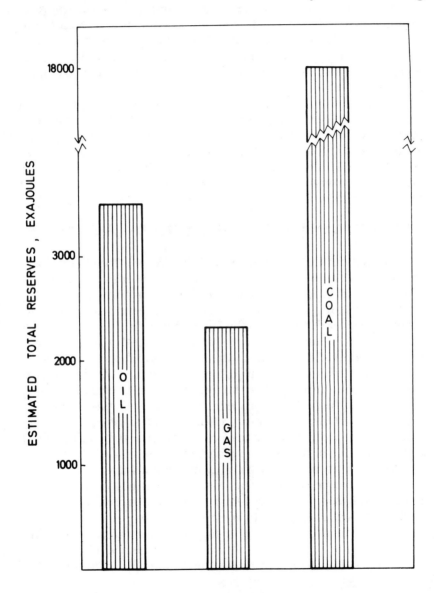

FIGURE 42. Reserves of nonrenewable forms of energy.

resources do not appear extensive and according to present estimates are only four times greater than the actual reserves. Future exploitation of petroleum will be more expensive because of poorer oil fields, which will require longer drilling and may be located in remote polar areas or at great depths in the ocean. The same holds for the exploration of oil shales and the tar sands.

Coal also provides a direct means of producing energy. It is easy to transport and store, but difficult and expensive to burn cleanly. The coal reserves are large enough for the next several hundred years. However, its combustion consumes oxygen and produces fairly large amounts of carbon dioxide and sulfur compound, together with nitrogen oxides, toxic metals, fly ash, organic carcinogens and mutagens, and natural radioactive substances.

The release of combustion gases into the atmosphere is a subject of concern even at great distances from the point of burning. Acid rains are presently causing severe damage to the environment and to the health of populations even on an international scale. The accumulation of carbon dioxide in the atmosphere may strongly affect the global climate in a not too distant future.

ENERGY FROM NUCLEAR FISSION — A REALITY

Throughout the world the end of the present decade has been marked by an antinuclear sentiment and its expression has varied in degree and manner according to the local tradition and social system. This antipathy arose even in countries in which achievements in nuclear energy initially found the favor of public opinion. Such was the case in France, where in 1991 nuclear power contributed to 72.7 percent of the total production in electricity.

Possible reasons for these feelings include the hazards to economy, environment, and health, and more recently, ethical and philosophical considerations relevant to long-term waste disposal. The Chernobyl catastrophe created a widespread belief that only a world power like the Soviet Union or the United States can technically handle such a disaster and economically survive.

Significant amounts of spent uranium fuel have been and still are accumulating since the last decades of nuclear plant operation. For disposal of this or of fission products from reprocessing, various procedures have been tried, but as yet not on a full industrial scale. Because of postponements and delays, the problem of nuclear reactor wastes and their disposal has become a heavy psychological burden for the public. The latter is tired of declarations that, compared to coal, the volume of nuclear waste is negligible (which is true), and that disposal is not a particular problem (which is rather far from being true).

It is, therefore, not surprising that the contribution of nuclear power to the world production of electricity was very modest during the 1980s and, most likely, will continue to be so during the last decade of this century. A "second nuclear era" seems to be needed if nuclear power is to make a significant and growing contribution to the generation of electricity. To again become publicly accepted, nuclear technology has to develop and test new reactor designs and demonstrate that they can produce economic and safe power. Additionally, the economics and safety of decommissioning power plants, and the subsequent disposal of wastes generated at all stages of production, must be demonstrated.

Despite all these drawbacks, nuclear energy is at the moment the only reasonable alternative to fossil fuel, and its contribution to the world production of electricity is undeniable. The International Atomic Energy Agency of the United Nations reports that by the end of 1991, 420 nuclear power plants were operating in 25 countries and providing about 12 percent of the world's electricity. The nuclear contribution in this respect was close to half or more of total electricity use in five countries: France (72.7%), Belgium (59.3%), Sweden (51.6%), Hungary (48.4%), and the Republic of Korea (47.5%).

The Thermal Neutron Reactor

Without oxygen consumption or emission of carbon dioxide, sulfur, or dust, this choice may be considered as an environmentally clean source of energy in comparison with the burning of fossil fuel. However, there are serious restrictions.

One of these is the accumulation of enormous amounts of radioactive materials in the reactor core and a subsequent potential danger in the event of natural catastrophic events or terrorist attacks.

The future protection of materials enriched in ^{235}U or ^{239}Pu may be particularly difficult. Their production is at present controlled by a few manufacturers, but an increase in the number of nuclear power plants or of installations for reactor fuel cycling will also increase the number of vulnerable sites. At the same time, the protection against loss or theft becomes more complex because of a higher frequency of transportation of fissile materials and of spent nuclear fuel.

The reserves of nuclear fuel for thermal neutron reactors are only moderate, in comparison with those of fossil fuel, and range between 1500 and 2500 exajoules (Figure 43). The resources of uranium are estimated at four times these amounts, not including that which is present in ocean water (Chapter 8).

In addition to the limited supply of uranium, the operation at safety of thermal nuclear reactors is also a major concern. Two accidents within only 7 years, Three Mile Island in 1979 and

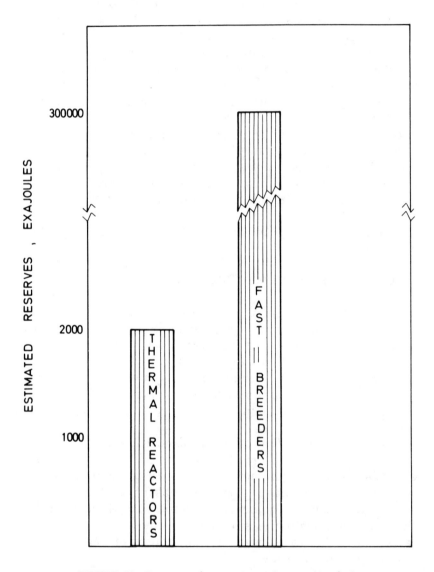

FIGURE 43. Reserves of energy in nuclear reactor fuel.

Chernobyl in 1986, were sufficient to establish the conviction that major nuclear accidents are not a trifling circumstance.

The nuclear industry in the West has tended to regard the accident at Chernobyl as a result of inadequate standards of Soviet engineering and quality control. This attitude rapidly caused a boomerang effect on domestic nuclear power programs. A meeting organized by the International Atomic Energy Agency on the Chernobyl incident showed that the differences between the Soviet and American reactors are without bearing on the cause of the disaster: it was a matter of human failure and of defective automatic regulation in the cooling of the reactor core. The fuel suffers irreparable damage and releases radioactivity if the core is uncovered or if the cooling water boils and turns to steam. Both types of reactors have a complex sequence of automatic emergency systems for keeping the core supplied with water. But there are some steps during normal operation, such as the shutting down of the reactor for maintenance, when the operators must switch off the automatic safety systems and intervene manually. The fatal error of both American and Soviet operators was that they misjudged the prevailing conditions in the reactor and shut off the emergency systems at the wrong moment.

The results of investigation on the two accidents have obliged the nuclear engineering community to replace the overconfident claim "it could not happen here" by the more realistic "it can also happen here", and the revisions and corrections of safety systems in many nuclear power plants have certainly since improved the overall safety factor. Nevertheless, the existing systems of security still rely both on human judgment and automation.

Inherent Safety

A new approach to reactor technology started to emerge in the early 1980s and was known as the concept of "inherent safety". Such security is not dependent on human intervention or on electromechanical devices, but on the immutable principles of physics and chemistry. It was incorporated in the Swedish project of the PIUS reactor, named from "Process Inherent Ultimately Safe".

Inherent safety is ensured by immersing the reactor core, the primary cooling system, and the steam generator in a large pool of water within a concrete vessel. The primary cooling system and the pool are hydraulically connected through pressure balanced interfaces placed at the top and bottom of the reactor. During normal operation, the hot water in the primary system is pumped through the core and the steam generators. The pressure developed by the coolant pumps is just enough to keep the pool water from entering the core. If for any reason the normal flow of primary cooling water is interrupted, the water from the pool immediately enters into the primary circuit through natural convection, floods the core, and cools it. Simultaneously, since it contains boron as an efficient neutron absorber it shuts down the chain fission process and prevents further generation of heat due to fission (Figure 44).

The system is designed so that the volume of water in the pool is sufficient to cool the reactor core for several days without the intervention of operators. This means that the protection is effective not only against potential accidents caused by human error or technical failure, but also against external events such as earthquakes, sabotage, or attack with conventional explosives.

There are other similar approaches to the inherent safety of nuclear reactors. Prior to construction and full-scale operation, however, the application of such precautions requires the solution of various delicate technical problems. In the case of the Swedish reactor, for example, the stability of the interfaces separating the primary coolant from the pool water is of primary importance. The maintenance procedures for this type of reactor will not be simple and remain to be devised.

The Fast Neutron Reactor

If nuclear power is to play a substantial role in the next century, it is certain that reactors with fast neutrons and breeders of fissile material offer interesting possibilities. The total energy reserves that the fast breeders could provide are estimated at 300,000 exajoules (Figure 43). The resources, not including uranium in the oceans, are estimated at 1 million exajoules. This means that fast breeder reactor power plants promise up to 50 times more energy than those which burn coal.

At present, experience accumulated by the operation of commercial-scale fast neutron reactors represents only a few tens of reactor-years, compared to more than four thousand reactor-years with thermal neutron reactors. Apart from several experimental units, there were in 1986 only three commercial-scale units in operation: one in the Soviet Union (550 megawatts of electricity) and two in France (270 and 1200 megawatts of electricity). The experience already acquired in the handling of plutonium and inflammable and radioactive sodium is still not enough to guarantee the operational success of technology based on fast neutrons.

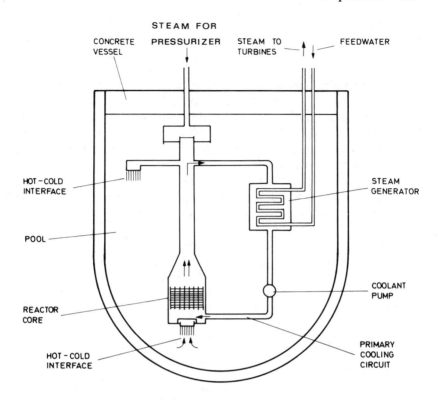

FIGURE 44. Scheme of PIUS, an inherently safe nuclear reactor (designed by ASEA-ATOM, Sweden).

Some data on the most ambitious fast neutron reactor project, the "Super Phenix" 1200 MWe power plant in France, may well illustrate the difficulties in the realization of such a commercial power unit. The reactor began operating in 1986, after 12 years of construction, and was in operation only one-third of the time until it closed in 1990. In fact, it only operated at full power (1200 MWe) for one month. From 1974, when the decision to construct was made, until it closed in 1990 about $5600 million was invested.

A main restraint in the future of fast breeders is also imposed by the plutonium fuel itself. Its amount is certainly not an obstacle, especially in the perspective of nuclear disarmament and the potential use of military stocks of plutonium. The major cause lies in the radiotoxicity of plutonium, and in the risk of reactor accidents leading to dispersion of the fuel. Also, when produced and used in massive quantities, there remains the factor of accidental plutonium criticality or of inadvertent release into the environment. It is clear that plutonium will also require "inherently safe" systems for its production and handling.

ENERGY FROM NUCLEAR FUSION — A MAJOR EVENT

Nuclear fusion is a reaction between light atomic nuclei which lead to formation of a heavier nucleus and the release of an important amount of energy. One way to fusion is penetrating the Coulomb barrier by force, by giving the nuclei a huge amount of kinetic energy. This usually requires heating the particles to tens of millions degrees. The process operates in the stars (Chapter 5) and was demonstrated on Earth with the explosion of the first thermonuclear device. The other way is to try to screen one nucleus from the repulsive effects of the other by binding the nuclei with a particle of opposite charge. The muonic hydrogen atom may be the right tool.

Research is continuing on a fairly large front, ranging from rather scarce studies on "cold" fusion catalyzed by muons at room temperature, to the widely supported activities which attempt to emulate stellar conditions with the use of extremely high-temperature plasmas or powerful lasers.

Muon and Cold Nuclear Fusion

The muon is a short-lived elementary particle with a half-life of 2.2 microseconds, which is similar to the electron but 207 times heavier (Table 4, Chapter 2). It appears in secondary cosmic rays and can be produced with high energy accelerators. Theoretical considerations indicate that because of its large mass, a negative muon can catalyze fusion in a mixture of deuterium and tritium. The process consists of a sequence of events governed by laws of atomic and molecular physics which do not require high temperatures. When exposed to muons, the nuclei of ^2H and ^3H fuse to form a single nucleus of ^5He, which breaks up into a free neutron and ^4He, i.e., an alpha-particle. The reaction releases energy as the neutron and α-particle move away from each other at high speed. The muon is left behind and is free to repeat its role as a catalyst, thereby maintaining a chain fusion reaction. However, this is not always the case, since the positively charged α-particle can capture the negative muon and interrupt the chain process. Experiments show that more than 100 fusion reactions per muon can be achieved, and different concepts of muon-catalyzed fusion processes have been proposed and investigated.

At present, many theoretical and experimental aspects still have to be clarified before a technical setup for cold fusion can be considered. Further advances in the technology of particle accelerators must also be made before the production of muons is economically feasible.

Plasma and High Temperature Fusion

Research is well advanced in the field of plasmas, which consist of a very hot, completely ionized gas. In a plasma, the numbers of positive ions and electrons are approximately equal and the material is virtually electrically neutral and highly conducting. It is considered as a fourth state of matter, in addition to the solid, liquid, and gaseous states.

Three basic requirements have to be satisfied for a sustained fusion reaction in plasma: the temperature must be high in order that the moving particles will gain energy and interact; the plasma density must be high enough for a sufficient number of reactions to occur, and the hot plasma must remain confined during the time required for reaction and the generation of energy.

Among several nuclear reactions of interest for fusion under terrestrial conditions, the most suitable is the fusion of ^{2}H, or deuterium, with ^{3}H, i.e., tritium. This reaction produces ^{4}He together with a neutron and a large amount of energy:

$$^2\text{H} + {}^3\text{He} \rightarrow {}^4\text{He} + \text{neutron} + 17.6 \text{ megaelectronvolts}$$

This fusion reaction also requires the lowest ignition temperature, which is about 40 million degrees kelvin, and its energy release per event is one of the highest.

The tritium consumed in the process is regenerated in a breeding blanket made of lithium, according to:

$$\text{Neutron} + {}^6\text{Li} \rightarrow {}^3\text{H} + {}^4\text{He}$$

$$\text{Neutron} + {}^7\text{Li} \rightarrow {}^3\text{H} + {}^4\text{He} + \text{neutron}$$

The breeding yield is about 1.3, i.e., up to 30 percent more tritium can be produced in the blanket than is consumed by fusion.

Approximately 80 percent of the energy produced in fusion is carried by the neutrons and appears eventually as heat in the lithium blanket. The latter consists of molten lithium metal, or an alloy of lithium and lead, or of a concentrated solution of a lithium salt. Heat is extracted by circulating the lithium blanket through a heat exchanger, which is used to produce steam for driving an electrical turbine. The remaining 20 percent of the energy released during the fusion process is carried by the helium nuclei and serves to maintain the temperature of the plasma.

The Break-Even Point

For industrial production, the energy extracted from a thermo-nuclear process must obviously be in excess of that required to heat and confine the plasma. The immediate goal of most labora-tory experiments is limited to a confinement of the plasma at or just above the ignition temperature until the energy released equals that required to produce and confine the plasma. There is a criterion known as the Lawson parameter, which is the product of the duration of confinement in seconds and the number of particles per cubic centimeter. In the deuterium-tritium reaction it is 3×10^{14} seconds per cubic centimeter, and can be achieved by maintaining the plasma of 3×10^{14} particles per cubic centime-ter for at least 1 second, or 3×10^{20} particles per cubic centimeter for 1 microsecond (10^{-6} second). The Lawson parameter is one of two that determine whether or not a fusion reaction will produce net positive energy. The other parameter is plasma temperature.

High temperatures and a close approach to the break-even point have been achieved with the experimental fusion assemblies known as tokamak. This device consists essentially of a doughnut shaped body (torus) in which a strong magnetic field provides confinement of the plasma; the name tokamak is derived from the Russian acronym for toroidal magnetic chamber (Figure 45). An extremely high vacuum and exceptional purity of the plasma are

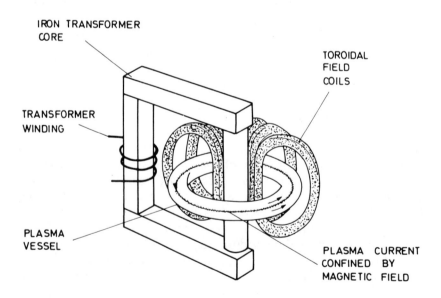

FIGURE 45. Scheme of magnetic confinement in a tokamak.

necessary to sustain fusion and minimize instability and loss of plasma. As an illustration, details are given for experiments with tokamak in 1986 at Princeton, New Jersey. The major radius of the torus is 2.5 meters and its minor radius 0.85 meter. The discharge current is 1 million amperes and lasts for 5 seconds. The current induces a magnetic field, which confines the high-temperature plasma and provides the necessary thermal isolation. It also heats the plasma, raising its temperature to 20 million degrees kelvin. Further heating, by injection of a stream of energetic neutral deuterium atoms into the plasma, increases the temperature at the center of the discharge to 230 million degrees kelvin. The ion density in the central region was 5×10^{13} particles per cubic centimeter, and the energy confinement time was between 0.1 and 0.2 second. In these tests the deuterium-tritium mixture was not used because the device was not equipped to handle radioactive tritium. Calculations show that if a 1:1 mixture of deuterium and tritium had been used under the same conditions, the generated fusion power would have exceeded 1 megawatt.

At year-end 1988, plasma temperatures of up to 350 million degrees, and fusion output powers up to 60 kilowatts, were reported. Generally, the experiments are meeting their objectives of elucidating the physics and engineering of the magnetic confinement of high-temperature plasmas in conditions close to those envisaged for commercial controlled-fusion reactors. However, the experiments have not achieved simultaneously the thermal insulation and high temperature needed for net energy output in the optimum fuel, a 50 percent mixture of deuterium and tritium.

Laser Fusion

The basic idea is to use high-energy pulsed beams of lasers to heat and ionize small pellets of fuel. The spherical pellet, of a few millimeters in diameter, is made of glass, plastic, or some other material and contains a mixture of deuterium and tritium. It is suspended in the target chamber, and a simultaneous discharge of several laser beams causes its implosion. Within a fraction of a picosecond (10^{-12} second) the implosion heats the fuel and increases its density by several orders of magnitude, thus creating propitious conditions for fusion. The highest density achieved at the end of the 1980s by the direct use of laser beams was 20 to 40 grams per cubic centimeter. However, it is necessary to compress the fuel of deuterium and tritium to much higher densities, typically to 1000 times their liquid density.

One suitable and powerful energy source is a neodymium-glass laser, which emits a single beam of infrared light at a wavelength of 1 micrometer. The beam is first amplified, i.e., its strength is increased by obtaining power from a source other than the input signal, and then it is split into a dozen separate beams. Each of these is first amplified and then its infrared light converted into ultraviolet radiation. The conversion to a shorter wavelength provides a more efficient energy transfer to the fuel pellet. Mirrors and lenses guide and focus the beams into the target chamber and direct the beam pulses toward the fuel pellet.

Within the past years, impressive energies of the order of 20 or 30 kilojoules per laser pulse have been produced, but this is still far less than the megajoules required for a fusion power reactor. Present-day lasers are still not very efficient in converting the energy which is fed into them; the efficiency of a neodymium glass laser is only about 0.2 percent.

In this respect, much better yields are provided by charged particle accelerators, which give up to 50 percent. All the same, progress in ion beam fusion is rather slow despite considerable financial support, which is not lacking owing to the potential interest in pulsed ion beams for weapons. A major difficulty is that accelerators have to provide high currents and high repetition rates. In addition, unlike lasers, accelerators cannot be scaled down for preliminary investigations. For these reasons, substantial research budgets are a prerequisite.

In a laser fusion plant power would be generated by microexplosions. Their frequency would depend on the amount of energy released per pellet, which might be as high as 100 megajoules. In this case, ten explosions per second would provide 1000 megawatts. This would be enough to operate a power plant of 300 megawatts of electricity. The reactor would also comprise a blanket of lithium to capture the neutrons released in thermonuclear reactions. The blanket would provide heat to generate steam and subsequently electricity (Figure 46).

An Appealing Goal

Several features make the fusion process an appealing source of energy. One is the fuel. Deuterium is present in seawater, with one molecule of D_2O for every 6500 H_2O, and can be extracted by fairly simple techniques. The tritium content in nature is negligible, but sufficient initial amounts can be produced by irradiation of lithium in a nuclear fission reactor. Further amounts of tritium can be produced in the lithium blanket of the reactor, and the breeder supplies sufficient quantities for operation of the fusion reactor.

A particularly important aspect of a fusion reactor is that the fusion process cannot get out of control as in the case of a fission

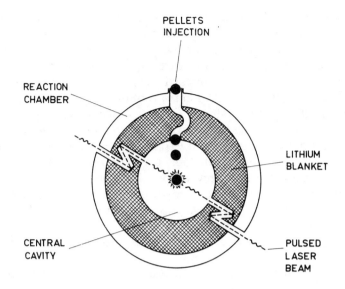

FIGURE 46. Scheme of inertial confinement fusion.

reactor. In the event of malfunction, the fusion process would simply cease and leave the power plant and environment untouched.

A relatively low degree of radioactivity accumulates in the fusion reactor because the main product of the process, helium, is not radioactive. The radioactive elements that are formed are due to nuclear reactions of construction materials with neutrons produced by fusion and which escape from reaction with the lithium blanket. Their activity will be significant, but small, compared to the amount of nuclear energy produced. An exception is the activity due to large amounts of tritium, both consumed and produced which will be unprecedented and will require very strict measures of radiation control and protection. About 500 grams of tritium will be consumed per day in a fusion reactor operating at 1200 megawatts.

A major obstacle lies in the complex and very costly technology. There are also questions of temperature, as in the provision of efficient supplementary heating of toroidal plasma in tokamak, or requirements for faster injection and irradiation of microspheres in a laser fusion reactor. Various problems are related to construction, as in the choice of a suitable reactor wall material for resisting the action of radiation, particularly neutrons. The appropriate form of lithium in the blanket is important because of the role of lithium as a coolant and the breeder of tritium.

The realization of hot plasma reactors requires very powerful magnetic fields capable of confining high-temperature plasma in volumes up to thousands of cubic meters. Superconducting magnet coils are thus of great importance for an efficient operation of fusion power reactors. Generally, the electrical resistance of a metal or alloy is a function of temperature. It decreases with temperature and tends toward zero as the temperature approaches absolute zero. For certain materials it becomes vanishingly small already a few degrees above absolute zero. This phenomenon is termed superconductivity and is used to produce large magnetic fields without expenditure of appreciable quantities of electrical energy. In fact, the only energy needed is that required to maintain the low temperature, which is achieved using liquid helium as coolant. Recent advances in the physics of superconductors which operate at the temperatures of liquid nitrogen or higher may revolutionize the construction of fusion power plants if the new superconducting materials can be designed to carry large currents.

The outlook for commercial viability of fusion power plants in the near future is still rather uncertain. Large investments in pilot setups are necessary before conclusions can be reached with respect to operation and economical feasibility on an industrial scale. The task is complex and costly, and international cooperation as in the case of the JET (Joint European Torus) is essential. The JET is presently the most powerful device in the world for studying fusion, and has been in operation since 1983 at Culham, near Oxford, England. It was built as a common project by countries of the European community.

Since 1987, a project to design what may be the first fusion reactor of the world is being considered under the auspices of the International Atomic Energy Agency. The United States, Soviet Union, Japan, and the European community are conjointly participating in the project ITER (International Thermonuclear Experimental Reactor). The goal is to produce a conceptual reactor design to test fusion technology by the end of the 1990s and to see how controlled fusion can operate in "star furnaces" on the Earth.

RADIATION AND MAN

From what we know or assume about the universe, radiation has been present throughout its evolution from the very beginning 1.5×10^{10} years ago, and will remain until its possible end some 10^{100} years after.

Our planet, and the life on it, are continuously bathed in radiations. In less than a century since the first discoveries were made, the understanding of radiation, radioactivity, and nuclear energy has made impressive progress. However, more has been accomplished in the mastering of complex technologies like those of nuclear power, than in the understanding of fundamental phenomena such as the nature of elementary constituents of matter, or the effects of radiation on life.

Man has still to learn how to live with radiation in its various forms. The main obstacle in the assessment of the nuclear future is most likely man himself. This is evident from the numerous impasses in the arms race and political issues relevant to nuclear energy. What Albert Einstein said 40 years ago may still be valid: "The unleashed power of the atom has changed everything save our modes of thinking, and we thus drift toward unparalleled catastrophes."

Perhaps the catastrophes are not inevitable. There exists an important, rapidly growing opposition to nuclear arms among the leading scientists of the world, and this sentiment is shared by a large part of the scientific community and the population as well. We are already witnessing the first steps toward the reduction of nuclear weapons which, hopefully, may lead to their complete elimination. The use of ionizing radiation and radionuclides, and the generation of nuclear electricity, are realities in our contemporary world, and we must live with them until better, nonnuclear solutions emerge. By then, effective international control of the radiation and radioactivity, in particular that of the release of radioactivity into the environment, should make living conditions more bearable in our global village.

Further Reading

Nuclear Power, the Environment and Man, International Atomic Energy Agency, Vienna, 1982. [A concise, clearly written survey of various aspects of nuclear power, the fuel cycle, nuclear wastes, nuclear safety, and radiation protection.]

Radiation, Doses, Effects, Risks, United Nations Environment Programme, Nairobi, Kenya, 1985. [Well illustrated, clearly presented, with short texts on radiation and life, sources of radiation, effects on man, and survey of units used in dosimetry.]

Ehrlich, P.R., Sagan, C., Kennedy, D., and Roberts, W.O., *The Cold and the Dark, the World after Nuclear War*, W.W. Norton, New York, 1984. [A classic work where, for the first time, the long-term, worldwide biological consequences of a nuclear war are considered. Presents the work reported at the conference on long-term consequences of a nuclear war. Despite the scientific style of presentation, the contributions are clearly written and can be read by nonprofessionals.]

Nuclear War, What's in It for You, Pocket Book, Simon & Schuster, New York, 1982. [A popular presentation of the nuclear arsenal, the ways it can be used, and the consequences. Honest selection of topics and exact presentation of facts. Should be read by everyone who wants information on nuclear weapons and their use.]

Morris, R., *The Fate of the Universe*, Play Boy Press, New York, 1982. [A physicist narrates the story of the universe in the light of more recent cosmological theories. A very clear, captivating description.]

Schell, J., *The Fate of the Earth*, Avon, New York, 1982. [A popular book in which the total impact of a major nuclear war is systematically analyzed.]

Taube, M., *Evolution of Matter and Energy on a Cosmic and Planetary Scale*, Springer-Verlag, New York, 1985. [Comprehensive for a large reading public, with a wealth of facts and numerical data.]

Glossary

Abberations, chromosomal: Abnormalities in number or structures of chromosomes*.

Abiotic: Refers to organic* matter which is produced chemically in the absence of living systems.

Acetylene: $HC \equiv CH$. A colorless, highly inflammable explosive gas. First member of the alkyne series, compounds containing the $C \equiv C$ group.

Adenine: $C_5H_3N_4NH_2$. A white crystalline substance. Occurs in nucleic* acids and plays an important role in the constitution of the genetic* code.

Aerobe: A cell or an organism that must use oxygen as the terminal electron acceptor for energy conversion.

Allene: $CH_2 = C = CH_2$. A colorless gas which tends to rearrange to the more stable isomer $CH_3C \equiv CH$.

Amino acid: An organic compound that contains both the carboxyl –COOH and the amino –NH_2 group, e.g., as in glycine: H_2NCH_2COOH. More than 80 amino acids are known, but only about 20 of them occur naturally in living organisms, in which they represent the building blocks of proteins*.

Ammonia: NH_3. A pungent gas that is very soluble in water and gives an alkaline solution of ammonium hydroxide, NH_4OH.

AMU or **a.m.u**: Abbreviation for "atomic mass unit", the unit of atomic weights in terms of an arbitrary standard (^{12}C).

---

* Terms with an asterisk are defined in the glossary.

Anaerobe: A cell or organism that cannot use oxygen for energy conversion and which must therefore employ some alternative energy-yielding process that is independent of oxygen.

Antibody: A chemical compound manufactured by the body to maintain immunity to infection by combining with and neutralizing harmful substances containing antigens*.

Antigen: A foreign substance (e.g., a protein*, a carbohydrate, or a virus) which enters into the biological processes of the body and stimulates the formation of antibodies*.

Angstrom, Å: Unit of length; $1Å = 10^{-10}$ m.

Archean: A period of the Earth's history from 4500 million to 2500 million years ago.

Asteroids: Planetoids, or miniature planets, which rotate around the sun in orbits between those of Mars and Jupiter. It is thought that there are many thousands of these bodies. Ceres, the largest of them, has a diameter of 1070 kilometers.

A.U., astronomical unit: A unit of length used in astronomy. It is the mean distance between the center of the Sun and the center of the Earth and is equal to 149.6 million kilometers.

Autotroph: An organism that uses CO_2 which is present in the environment, or generated from some other compound, as the immediate (and major) source of cellular carbon.

Bacteria: Class of microscopic, unicellular, prokaryotic* organisms which vary widely in size, form, and characteristics of growth. They can derive energy from a large number of sources.

Banded iron formation: Chemical sedimentary rock with alternating layers of reduced and oxidized iron minerals. They provide evidence for the introduction of oxygen into the Earth's atmosphere.

Biogenic: Of biological origin; formed by the activity of organisms.

Biosphere: General term for regions in which life can exist.

Black body radiation: Radiation of all frequencies, such as would be emitted by an ideal "black body" which absorbs all radiation falling upon it.

Blue-green algae: Prokaryotic*, bacterium-like microorganisms; mainly aquatic plants whose chlorophyll contains bluish-green pigments. Also called cyanobacteria. Capable of aerobic (oxygen-producing) photosynthesis; numerous strains are also capable of anaerobic (nonoxygen-producing) photosynthesis. They first appeared 2200 to 3500 million years ago; their life processes served to add oxygen to the atmosphere and create the stromatolites*.

Bond: The linkage between atoms in molecules and between molecules or ions in crystals.

Bond energy: The energy characterizing a chemical bond between two atoms. Measured by the energy required to break that bond, e.g., 348 kilojoules per mole for C–C; the energy is generally greater for multiple bonds (837 kilojoules per mole for C≡C).

Carbonaceous: Refers to an object or substance that contains organic matter.

Carbon dioxide: CO_2. A colorless gas, heavier than air, which occurs in the atmosphere as a result of the oxidation of carbon.

Carbon monoxide: CO. A colorless, very poisonous gas that is formed during the incomplete combustion of carbonaceous materials.

Carboxylic acid: Any of the many organic acids containing one or more carboxyl (–COOH) groups.

Carcinogen: A substance capable of causing the development of malignant cells.

Catalyst: A substance that increases the rate of a chemical reaction without itself appearing among the products of the reaction. Most of the chemical reactions in living organisms are catalyzed by certain protein* molecules which are called enzymes*.

Chain reaction: Any self-sustaining molecular or nuclear reaction, the product of which contributes to the propagation of the reaction.

Chromosomes: Thread-like bodies that carry the genetic material; they occur in the nuclei of eukaryotic* cells.

Cell: All living organisms are composed of discrete, membrane-enclosed biological units. Many microorganisms (e.g., bacteria) consist of only one cell, whereas more complex organisms are composed of a multitude of cooperating cells.

Chemical sediment: Sediment or sedimentary rock composed primarily of material formed by precipitation from solution or colloidal suspension, often as a result of evaporation. They exhibit a typical texture, as in limestones (natural calcium carbonate, $CaCO_3$).

Chondrite: A stony meteorite that is characterized by the presence of chondrules, spheroidal granules about 1 millimeter in diameter consisting usually of the mineral olivine, a silicate containing magnesium and iron.

Core of the Earth: The spherical central zone of the Earth's interior, with an overall radius of 3471 kilometers ; it consists of a solid inner core of radius 1291 kilometers and a liquid outer shell of 2180 kilometers in thickness.

Crust of the Earth: The outermost shell of the planet (less than 0.1 percent of the Earth's total volume). It extends to a depth of 30 to 50 kilometers beneath most continents and about 10 to 12 kilometers beneath the oceans.

Crystalline: Refers to substances consisting of crystals* or solidified in the form of crystals.

Crystal: Substance that solidifies in a definite geometrical form. Because of the tendency of the molecules to attain the lowest potential energy state, the molecules tend to pack as loosely as possible. Crystals are classified according to the geometrical patterns they assume in the crystal lattice or to the type of interatomic bond that holds them together. Crystals may be ionic (sodium chloride, NaCl), covalent (diamond, silicon, most organics), or metallic (most metals).

Cyanamide: H_2NCN. A colorless, crystalline, unstable substance.

Cyanides: Salts of hydrogen cyanide (e.g., KCN, NaCN)

Cyanopolyyne: $HC \equiv C - C \equiv C - C \equiv N$. Acetylenic hydrocarbons with a cyano group ($-C \equiv N$).

Cytoplasm: The part of the cell* surrounding the nucleus and bounded by an intracellular (cytoplasmic) membrane.

Defect: A discontinuity in the pattern of atoms, ions, or electrons in a crystal.

Deoxyribonucleic acid, DNA: Long thread-like molecules consisting of two interwound helical chains of polynucleotides. The sugar of all nucleotides* is 2-deoxy-D-ribose, but each nucleotide is characterized by one of the four following nitrogenous bases: adenine, cytosine, guanine, and thymine. DNA molecules are responsible for storing the genetic* code by the order of the arrangement of their nitrogenous bases, three bases coding for one amino acid*. The structure of a DNA molecule can be compared to a twisted rope ladder, the sides of which consist of sugar-phosphate chains and the rungs of hydrogen-bonded*, nitrogenous bases.

Deoxyribose: $C_5H_{10}O_4$. A colorless, crystalline sugar consisting of five carbon atoms.

Dissociation: Reversible decomposition of molecules which occurs under particular conditions. In electrolytic dissociation the molecule separates into ions.

Enzyme: A large group of proteins produced by living cells, and which act as catalysts* in biochemical reactions. They are effective in minute quantities and are highly specific in their action.

Eon: Any large part of geological time.

Esters: Organic compounds corresponding to inorganic salts. Formed by the chemical combination of an acid and an alcohol with elimination of water.

Ethane: CH_3CH_3. A colorless, odorless gas.

Ethanol: CH_3CH_2OH. Ethyl alcohol, a colorless, inflammable liquid.

Ethylene: $CH_2 = CH_2$. A highly reactive, unsaturated hydrocarbon.

Ethyl cyanide: CH_3CH_2CN. An alkyl cyanide also called propionitrile.

Eukaryote: Refers to cells whose internal construction is complex, consisting of organelles* (e.g. nucleus), chromosomes*, and other structures. All higher organisms and many unicellular organisms are composed of eukaryotic cells. Apparently, the evolution of complex life was subsequent to that of the eukaryotic cells, an event that occurred about 1000 million years ago.

Exobiology: Extraterrestrial biology is the study of life forms as they might occur outside the terrestrial environment.

Exogenic: An externally acting geologic process such as weathering or erosion.

Formaldehyde: HCHO. The simplest aldehyde. Colorless gas with a pungent odor.

Formamide: $HCONH_2$. An odorless, colorless liquid.

Formic acid: HCOOH. The simplest monocarboxylic acid. A colorless, corrosive liquid with a penetrating odor.

Fossil: Any direct physical (viz., morphological fossil) or chemical (viz., chemical fossil) remains, object, or structure which is indicative of preexistent life.

Genetic code: The code by which inherited characteristics are handed down from generation to generation. The code is expressed by the molecular configuration of the cell constituents called the chromosomes*.

Giga: Prefix signifying 10^9, symbol G.

Gigawatt of electricity: Unit of electric power equal to 10^9 watts, GW(e).

Glucose: $C_6H_{12}O_6$. A sugar containing six carbon atoms (hexose).

Glycine: H_2NCH_2COOH. The simplest amino acid*.

Glyoxal: CHOCHO. A yellow, crystalline, organic compound.

Glyoxylic acid: CHOCOOH. A syrupy organic acid, which also reacts as an aldehyde.

Gonads: The ovaries or testes.

Guanidine: $HNC(NH_2)_2$. A strongly basic, crystalline, organic compound.

Heterotroph: An organism that uses organic carbon compounds as sources of cellular carbon.

Hydrocarbons: Compounds composed only of hydrogen and carbon.

Hydrosphere: The aqueous envelope of the Earth, including ocean water and vapor in the atmosphere.

Hydrogen cyanide, hydrocyanic acid: HCN. A highly reactive, very poisonous compound. It is colorless, burns in air, and has an odor of bitter almonds.

Hydrogen bond: A weak electrostatic bond that is formed between a hydrogen atom and a strongly electronegative atom (e.g., oxygen, nitrogen, fluorine). The molecules in liquid water or in ice are held together by hydrogen bonds (O–H—O). The hydrogen bond is of paramount importance in biochemical processes, especially the N–H——N bond, which participates in the formation of proteins* and nucleic* acids. Life as we know it would be impossible without this type of chemical linkage.

Hydrogen ion: H^+. A positively charged hydrogen atom or proton. The general properties of acids in solution are due to the presence of hydrogen ions. Their actual form in water is the hydronium ion H_3O^+.

Hydrogen sulfide: H_2S. A colorless, poisonous gas with a smell of bad eggs. It is formed by the decomposition of organic compounds containing sulfur.

Igneous: Refers to a rock or mineral that has solidified from molten or partly molten material (from a magma*).

Inorganic: Of mineral origin, not belonging to the large class of carbon compounds that are termed organic.

Kilo: Prefix denoting 10^3, symbol k.

Kerogen: Organic material lacking a regular chemical structure, insoluble in organic solvents and mineral acids. Present in sedimentary rocks as a result of geochemical alterations.

Leukemias: Forms of cancer due to multiplication of cells of bone marrow.

Light-year: An astronomical measure of distance; the distance traveled by light in 1 year, equal to 9.4605×10^{12} kilometers.

Limestone: Chemical sedimentary rock composed chiefly of calcium carbonate minerals.

Lithosphere: The solid part of the Earth as opposed to the hydrosphere* and atmosphere; also, the crust and the upper portion of the outer mantle* of the Earth.

Magnetosphere: The region of the magnetic field surrounding the Earth, extending out thousands of kilometers.

Magma: Molten or semimolten rock material.

Mantle of the Earth: The shell of the Earth below the crust and above the core. The upper mantle extends to a depth of 900 kilometers, and the lower mantle to a depth of 2900 kilometers.

Mega: Prefix denoting 10^6, symbol M, e.g. Mrad = 10^6 rad.

Metabolism: Chemical processes associated with living organisms. The metabolism of a substance refers to the changes that the substance undergoes in an animal or plant.

Metamorphism: The mineralogical and structural conversion of rocks under physical and chemical conditions that prevail at a certain depth in the Earth's crust, and which differ from the conditions under which the rock was originally formed.

Methane: CH_4. An odorless, inflammable gas that forms an explosive mixture with air.

Methanol: CH_3OH. Methyl alcohol.

Methyl amine: CH_3NH_2. A gas with an odor resembling that of ammonia.

Methyl cyanide: CH_3CN. A colorless, poisonous liquid, also called acetonitrile.

Micron or micrometer: A measure of length, 1 micrometer = 10^{-6} meters.

Nanometer: A measure of length, 1 nanometer = 10^{-9} meters.

Nitrile: Organic compounds containing the functional group $C{\equiv}N$. On hydrolysis they yield a carboxylic* acid containing the same number of carbon atoms as the nitrile.

Nonstochastic radiation effects: Those for which the severity of the effect varies with the magnitude of the dose causing it, and for which the likelihood of the effect is high once a threshold dose is exceeded.

Nucleic acids: Long, chain-like molecules which, in the various combinations of the constituent groups, embody the genetic* code (deoxyribonucleic* acid, DNA*) and serve in its transmission (ribonucleic* acid, RNA).

Nucleoside: A compound formed from a nitrogenous base (purine* or pyrimidine*) and a pentose* sugar.

Nucleotide: An important type of compound found in all living matter. It is the unit component in the structure of DNA*, and is repeated every few nanometers along the length of the DNA molecule. Consists of a nitrogenous base (purine* or pyrimidine*), a pentose* sugar, and a phosphate* group.

Organ and tissue: The meanings overlap. The term "organ" applies when all body cells of a certain type occur together in the same structure, as in liver or brain. "Tissue" applies when cells of a certain type occur in various parts of the body, e.g., those of the bone marrow or lymph glands. Radiation affects cells of most of the organs and tissues of the body.

Organelle: A specific cellular part analogous to an organ.

Organic: A term relating to a compound, structure, or substance containing carbon and usually hydrogen, oxygen, and/or nitrogen, of the type characteristic of biological systems; there also exists abiotic* organic material.

Oxidation: The combination of oxygen with a substance. More generally any reaction in which an atom or ion loses electrons, e.g., the change of a ferrous ion (Fe^{++}) to a ferric ion (Fe^{+++}).

Oxidizing agent: A substance that causes an oxidation reaction to occur.

Oxidation-reduction or a redox reaction: A chemical reaction in which an oxidizing* agent is reduced and a reducing* agent is oxidized, thus involving the transfer of electrons from one atom, ion, or molecule to another.

Oxalic acid: HOOCCOOH. A white, crystalline organic compound.

Ozone: O_3. A molecule containing three atoms of oxygen. It is a bluish gas, chemically very active and a powerful oxidizing agent. Ozone in the atmosphere is mainly present as the ozone layer, at 30 to 45 kilometers above the Earth's surface.

Pasteurization: Partial sterilization in which bacteria, but not bacterial spores, are killed.

Parsec: A unit of length used in astronomy and equal to 3.084×10^{13} kilometers or 3.26 light-years, symbol p.

Pentose: A sugar with five carbon atoms. The most important is ribose, an essential constituent of nucleic* acids.

Peptide: A compound of two or more amino acids* formed by combination of the amino group ($-NH_2$) of one acid and the carboxyl group ($-COOH$) of another. It is characterized by the peptide linkage ($-HN-CO-$).

Phosphate: A salt of phosphoric acid (H_3PO_4).

Physicochemical: Caused by the action of, or involving the principles of both physics and chemistry.

Photolysis: The decomposition or reaction of one or more substances on exposure to light.

Photosynthesis: Process by which the green plants manufacture their carbohydrates from atmospheric carbon dioxide and water in the presence of light.

Photochemical reaction: Chemical reactions which are initiated, assisted, or accelerated by exposure to light.

Photoautotroph: Autotrophic* and deriving energy from light.

Polymerization: Process of forming long molecules (polymers) starting with small constituent units (monomers).

Polypeptide: A chain comprising three or more (up to several hundred) amino acid* units joined by peptide* linkages.

Precambrian: The earliest eon* of the Earth's history. It includes the Archean* (4500 to 2500 million years ago) and the Proterozoic (2500 to 570 million years ago) periods.

Prokaryote: Microorganism characterized by cells that lack membrane-bonded nuclei, organelles, and other complex constituents, like bacteria* or cyanobacteria (blue-green algae*).

Protoplasm: The matter of which biological cells consist.

Proteins: Large organic polymers which consist of polypeptide* chains cross-linked in various ways. They are essential constituents of all living cells.

Purines: Organic compounds of general formula $C_5H_4N_4$, which form with sugar and phosphate a component of nucleotides* and nucleic* acids.

Pyrimidines: Organic compounds of general formula $C_4H_4N_2$, which are constituents of nucleotides* and nucleic* acids.

Reduction: The removal of oxygen from a substance or the addition of a hydrogen atom to it. The term is also used more generally to include any reaction in which an atom or ion gains one or more electrons.

Reducing agent: A substance that removes oxygen from another substance. More generally, a substance which donates electrons.

Ribonucleic acid (RNA): The molecular structure is the same as in DNA* except that RNA contains the sugar ribose instead of deoxyribose* and the base uracil in place of thymine.

Ribosomes: Complex structures within the cytoplasm of cells which serve to synthesize proteins*.

Sandstone: A rock formed from sand or quartz particles cemented together with clay, calcium carbonate, and iron oxide.

Sedimentation: The process of settling out an insoluble solid from a liquid in which it is suspended.

Shock wave: Discontinuity in the flow of a fluid (including a gas or a plasma), marked by an abrupt increase in pressure, temperature, and flow velocity at the shock front.

Sulfur dioxide: SO_2. A colorless gas with a penetrating odor.

Somatic effects: Those expressed in body (somatic) organs or tissues, as distinct from genetic effects expressed in later generations, and due to action on gonads*.

Stromatolite: A laminated, accretionary, organo-sedimentary fossil structure produced by the growth and metabolic activities of simple organisms (usually cyanobacteria). Common in Precambrian times, much rarer today.

Stochastic effects: Those for which the probability of occurrence, rather than the severity of an effect, varies with the size of the dose. For such effects (e.g., the induction of genetic defects or cancer), no threshold dose can be assumed below which some deleterious consequence may not occur.

Tera: Prefix signifying 10^{12}, symbol T.

Index